RAND 1700 Main Street, PO Box 2138, Santa Monica, CA 90407-2138

PUBLICATIONS DEPARTMENT

ERRATA

February 11, 2000

To: Recipient of MR-1100-OSD

Title: *Swarming on the Battlefield: Past, Present, and Future*

Author: Sean J. A. Edwards

From: Publications Department

Please note the following correction to Page 82, Table 5.1, under the column heading "Napoleonic Wars." The entry 4.79 should read 4,790. We apologize for the confusion.

SWARMING
ON THE
BATTLEFIELD

Past, Present, and Future

SEAN J. A. EDWARDS

Prepared for the
Office of the Secretary of Defense

National Defense Research Institute

RAND

PREFACE

The military application of emerging technologies for communications and information processing is likely to change the way military force is managed and applied. Oftentimes, a dramatic improvement in technology brings about an equally dramatic change in military doctrine and organization. One possibility is a doctrine based on swarming, whereby military units organized as networks use dispersed yet integrated operations.

Swarming has occurred throughout military history, and the lessons of this past experience may offer insights into a possible future application of swarming. Very little historical research has been conducted on the use of swarming, let alone a comprehensive review of swarming as a major theme within military history. This monograph seeks to address this deficiency by analyzing ten swarming cases. The conclusions of this historical analysis are then applied to a discussion of future swarming. This work should be of interest to U.S. policymakers, commanders, planners, and others who desire an understanding of swarming and its potential for future light operations or as a new way of warfare.

This study was prepared for the project "Swarming and the Future of Conflict," which was directed by John Arquilla and David Ronfeldt. The project was sponsored by the Office of the Assistant Secretary of Defense for Command, Control, Communications and Intelligence (C3I) and conducted in the International Security and Defense Policy Center of RAND's National Defense Research Institute, a federally funded research and development center sponsored by the Office of the Secretary of Defense, the Joint Staff, the unified commands, and the defense agencies.

CONTENTS

FIGURES

TABLES

The future of war is fraught with uncertainty. Whether or not a revolution in military affairs (RMA) is about to occur, the rapid pace of technological innovation is dramatically enhancing the ability to see, track, and kill targets on the battlefield.[1] At the same time, the proliferation of weapons of mass destruction (WMD) and the increasing lethality of cluster and precision guided munitions make it imperative that future ground forces remain dispersed as much as possible. This requirement for dispersion begs the question of whether the traditional hierarchical organization and doctrine of the U.S. Army should change.

Current discussion about future doctrine for U.S. ground forces already involves concepts such as dispersed operations, networking, and greater autonomy for small units than has been customary. One important part of that doctrinal discussion relates to the feasibility and utility of "swarm tactics," which would have such small, distributed units and maneuverable fires converge rapidly on particular targets. To help inform the debate over the potential relevance of swarming to U.S. military doctrine, this monograph analyzes swarming examples throughout military history. A close reading of those examples might reveal historical patterns and lessons that remain important today.

---

[1]An RMA occurs when the application of new technologies combines with innovative operational concepts and organizational change in a way that fundamentally alters the character and conduct of conflict.

APPROACH

It is beyond the scope of this monograph to look at every swarming example in military history. Instead, the analysis uses a cross section of cases, each case having one or more unique critical characteristics. The goal is to cover the prototypical cases, not to exhaustively describe every case that ever occurred. Several research questions are explored, including the following:

- Are there any apparently dominant factors?

- How well do swarmers do against nonswarmers?

- Does swarming have particular advantages for offense rather than defense, or does it apply to both?

- How does swarming success vary according to terrain?

- How did swarmers satisfy their logistics demands?

- What have been the successful countermeasures to swarming?

For the purposes of this monograph, a *swarming case* is any historical example in which the scheme of maneuver involved the convergent attack of several (or more) semiautonomous (or autonomous) units on a target force in some particular place. The following list presents ten general cases between a swarming and a nonswarming force that are examined and compared in the monograph. For each of these cases, a preliminary discussion of the belligerents is followed by a focus on one or more particular battles. The general cases are listed below (assume year A.D. unless otherwise noted):

- Scythians versus Macedonians, Central Asian campaign including the Battle of Alexandria Eschate, 329–327 B.C.

- Parthians versus Romans, Battle of Carrhae, 53 B.C.

- Seljuk Turks versus Byzantines, Battle of Manzikert, 1071

- Turks versus Crusaders, Battle of Dorylaeum, 1097

- Mongols versus Eastern Europeans, Battle of Liegnitz, 1241

- Woodland Indians versus U.S. Army, St. Clair's Defeat, 1791

- Napoleonic Corps versus Austrians, Ulm Campaign, 1805

- Boers versus British, Battle of Majuba Hill, 1881
- German U-boats versus British convoys, Battle of the Atlantic, 1939–1945
- Somali insurgents versus U.S. Commandos, Battle of the Black Sea, 1993.

The cases cover two types of identified swarming maneuvers, various combinations of opposing-force types, and scale and spectrum of war.

Most examples of military swarming are tactical "Massed Swarm" cases from the ancient world and the Middle Ages, wherein a swarming army begins as a single massed body then disassembles and conducts a convergent attack.[2] Other swarming examples are "Dispersed Swarm" cases, such as those drawn from the history of guerrilla warfare, wherein the swarming army is initially dispersed but then converges on the battlefield without ever forming a single mass.

HISTORICAL FINDINGS

Across the identified cases, at least three factors appear to play a role in whether or not swarming is successful: elusiveness (either through mobility or concealment), a longer range of firepower (standoff capability), and superior situational awareness. The actual source of these key advantages varied from case to case. Simple pattern-matching reveals that elusiveness and situational awareness appear to be more important than standoff capability. The combination of these three key advantages appears to have a synergistic effect.

Swarming offers several tactical advantages, including the following:

- For a swarming army to attack a defender from all sides appears to have an unnerving psychological effect, and it creates killing zones.

---

[2]In earlier swarming cases, command, control, and communications (C3) was too primitive to allow greater massing or dispersion. Without the benefit of wireless communication, it was difficult to coordinate many units without keeping them within sight of each other.

- Deceptive swarmer tactics such as feigned retreats and ambushes are very successful against undisciplined opponents.

- It can sever the nonswarming army's lines of communication.

- Networks are better at fighting other networks.[3]

- It gives the ability to choose the time and place of battle.

Swarming has not always worked, however. The historical record indicates several countermeasures to swarming. Swarmers could be defeated if their mobility advantage was negated. Past examples of this countermeasure include pinning more-elusive swarmers against a geographic obstacle such as a river or a fort, or using a friendly unit as bait to lure the swarmer into a trap. Terrain such as mountain passes and other obstacles were used to channel swarming armies and hinder their mobility. In other cases, the loss of standoff fire capability defeated the swarmer. Finally, attacking the logistics vulnerabilities of swarmers also proved successful in some cases.

Successful countermeasures to swarming highlight the limitations of and constraints on a swarming doctrine. Some of the historical limitations to swarming include the following:

- Swarmers were sometimes incapable of a quick knockout blow.

- The mobility and/or concealment of past swarmers depended on the terrain. Various swarmers have relied on heavy woodlands, urban areas, oceans, and grasslands capable of supporting many horses.

- Logistics was a significant challenge. Even when insurgent swarmers were able to rely on the support of indigenous populations and the countryside, they rarely fielded major forces for any sustained campaigning. For the operational-level swarmers such as the Mongols and *La Grande Armée*, a logistics breakthrough was necessary.

- Swarming attacks on fixed defenses yielded mixed results. The siege trains necessary to breach heavily fortified towns and cities

---

[3]This point is argued in John Arquilla and David Ronfeldt, *The Advent of Netwar*, Santa Monica, CA: RAND, MR-789-OSD, 1996.

generally were available only after a decision was reached in the field. Among the horse-archer cases, only the Mongols were successful at reducing strongholds, because of their Chinese siege-craft engineers and the calculated use of terror.

IS A SWARMING DOCTRINE FEASIBLE?

Ultimately, a swarming doctrine's feasibility will depend on the benefits emerging from the information and communication revolutions. Many benefits are already being realized. Eventually, the computers and wireless radios that are currently being installed on every vehicle, plane, and ship will communicate digitally with each other across interoperable battle command and control systems. Sensors and shooters will share a near-real-time common picture of the battlefield. Several governmental research and development efforts under way may be relevant to a discussion of swarming, including the U.S. Army's Army XXI and Army After Next (AAN) work and the Marine Corps' Urban Warrior program.

Even though no major military power has ever tried swarming on land in the modern age of mechanized war, many of the historical limitations and countermeasures from the past appear tractable given technological or doctrinal development in the future. For example, the occasional limitation of swarmers in the past—an inability to deliver a knockout blow quickly—may be overcome in the future by joint, indirect fire assets such as fighter/bombers, C-130 and helicopter gunships, the multiple-launch rocket system (MLRS)–fired Army Tactical Missile System (ATACMS), offshore Naval fire support, and even space-based kinetic-energy weapons. The command and control limitations of the past may be erased by the advent of future wireless communication systems—such as mobile mesh networks—capable of supporting a tactical internet anywhere in the world.

Given the radical force-structure changes a swarming doctrine would require, this study recommends that a portion of the U.S. light or medium force adopt swarming as an operational concept, if swarming proves to be feasible during field experiments. History suggests that swarming armies were successful when they were able to elude their opponent, possessed standoff firepower, and enjoyed superior situational awareness. It is reasonable to assume that swarming can work again if future forces enjoy these same advantages. Ongoing technological development suggests that light Army units may soon enjoy them.

Since a future swarming doctrine is still very much a concept in progress, additional detail is offered here on the tactics, logistics, command, and organization of a possible swarming doctrine. This speculative discussion is based as much as possible on the historical conclusions from the ten cases.

Tactically, swarming can be conceptually broken into four stages: locate, converge, attack, and disperse. Swarming forces must be capable of *sustainable pulsing*, coalescing rapidly and stealthily on a target, then redispersing and recombining for a new pulse. Because of the increasing vulnerability of massed formations on the ground to airpower and WMD, the Dispersed Swarm maneuver is more appropriate than the Massed Swarm maneuver for the future. Operations that are more dispersed are a natural response to the growing lethality of modern munitions.

The logistics problem of supplying widely dispersed units is a difficult one and will probably need to be addressed by a package of fixes, such as reducing the demand for fuel and ammunition by using lighter vehicles and more indirect-fire assets; streamlining the logistics process with information-processing techniques such as focused logistics;[4] using precise aerial resupply when possible; and prepositioning supply depots.

As to command, a swarming doctrine must take into account the coordination of many small units. To be sure, the complexity of the

---

[4]Focused logistics uses a Velocity Management approach to battlefield distribution, wherein the speed and control of logistics material is more important than the mass of stockpiles. By re-engineering logistics processes, Velocity Management can reduce the long material flows that help create massive stocks of supplies. Eliminating non–value-added activity and maintaining in-transit visibility (or knowing where every logistics item is at all times) decreases the logistician's response time to warfighter demands. In the past, U.S. inventories have typically been large because warfighters hoarded supplies "just in case" the items they ordered either took too long to arrive or never showed up. Rather than "just-in-case," focused logistics seeks to respond to real-time battlefield demand and move in the direction of a "just-in-time" philosophy. Rapid response to the needs of dispersed maneuver units will provide logistics support in hours and days rather than weeks.

command grows with the number of units, the power and range of their weapons, the speed at which they move, and the space over which they operate. To handle this complexity, both the supply and demand of command must be addressed. Improved command, control, communication, computers, and intelligence (C4I) technologies may provide part of the answer by increasing the supply of command, but a doctrine based on swarming will have to provide other ways to reduce the demand for command. One way to coordinate many small units is to use a decentralized network organization. A decentralized network calls for semiautonomous units to follow the mission-order system of command—what the Germans called *Auftragstaktik*—whereby small-unit commanders have the freedom to deal with the local tactical situation on the spot while following the overall commander's intent. Swarming would be difficult for a hierarchical command structure, because the extremely flat organization of a dispersed network would place too much demand on mid-level and higher commanders.

The extent to which a future swarming doctrine depends on superior technology is a key question. History demonstrates that technological advantage is *always* temporary. That said, there are several key functions that new weapons and technology must provide. Real-time situational awareness will require the integration of command and control systems, communications systems, and intelligence, surveillance, and reconnaissance (ISR) systems. The communications system for a dispersed tactical formation would have to be a mobile mesh communications network with high data throughput and survivability. Rapidly responsive indirect precision fires delivered by rockets, missiles, naval gunfire, or tactical air must be available for swarming to deliver enough firepower. And a next-generation Light Strike Vehicle will be needed to provide swarming units the mobility they need to remain elusive.

Battles are won by the careful meshing of one side's advantages with the other side's weaknesses. Swarming is no exception. As with any tactic or strategy, swarming will not work against all types of opponents in all situations. It is posited here that network organizations that use a dispersed formation to swarm are more relevant than AirLand Battle divisions for four particular missions:

- Power-projection missions

- Dispersed operations

- Counterinsurgency operations

- Peace operations.

The Army recognizes the need for a light force that can deploy rapidly and halt enemy armored movements. The mobile yet lethal nature of the swarm unit lends itself well to the missions of light expeditionary forces—such as the "Halt Phase" mission[5]—because swarm units are light enough to be air-transported and mobile enough to remain elusive. Today's traditional light airborne and air assault units are composed of dismounted infantry and are incapable of stopping enemy armor threats.

The dispersion of swarming units on the battlefield will reduce their vulnerability to weapons of mass destruction. In addition, if adversaries of the United States adopt dispersion themselves as a tactical countermeasure to the increasing lethality of U.S. air-delivered weapons, swarm units will be needed to flush out the enemy and facilitate target acquisition for U.S. air forces.

A network of swarm units dispersed over an area can perform counterinsurgency (COIN) missions well, because a highly mobile network of nodes can detect dismounted enemy personnel more effectively than can standard U.S. reconnaissance assets. A swarm force can physically "cover down"[6] over a geographic area and pick up battlefield intelligence missed by friendly airborne and spaceborne sensors.

Swarm units may be more effective at performing peace operations than are traditional combat divisions. They have the advantage over hierarchical divisions when organized for peace operations, an

---

[5]A Halt Phase mission is aimed at halting an enemy invasion short of its objectives in major theater wars. During the initial stages of an enemy armored offensive, the only U.S. forces available are joint forces deployed in the theater during peacetime, air forces, and rapidly deployable light forces. The heavy ground forces that are needed to evict enemy forces from captured territory must be sea-lifted to the theater, a lengthy process that takes at least several weeks.

[6]To *cover down* is to blanket or cover an area with numerous personnel. Units physically deploy in enough local areas so that no area is left uninvestigated.

ability to shape the environment, and an ability to minimize command problems on urban terrain.

A FINAL NOTE

Swarming is not new. During the pre-gunpowder age (i.e., before the fourteenth century), swarming armies enjoyed quite a bit of success on the Eurasian steppe and elsewhere; more recently, light infantry insurgents have fared well against conventional armies. The question is, Does a role exist for swarming today or in the future? History strongly suggests the answer is yes, if three capabilities can be achieved: superior situational awareness, elusiveness, and standoff fire. If emerging technology provides these capabilities, the U.S. military could enter a watershed era of modern swarming that involves dispersed but integrated operations. Any doctrine of the future that relies on dispersed operations, such as AAN or Urban Warrior, could benefit from a sustained research effort on swarming.

The patterns and conclusions presented in this study are preliminary. They are based on a carefully chosen yet limited sample; further research on additional cases would help validate or complete the analysis begun here. Many other historical candidates remain, both from other battles between the belligerents surveyed in this study and possible new cases such as the Battle of Britain in 1940; the defensive *Luftwaffe* tactics used over Germany late in World War II; the Chinese infantry tactics used in the Korean War, 1950–1952; the North American Indian Wars of the nineteenth century; and, more recently, the ongoing guerrilla war in southern Lebanon. A closer look at battles between swarmers themselves, such as at the thirteenth-century Battle of Ayn Jalut in what is modern-day Syria between Egyptian Mamluks and the Mongols, would explore how elusive forces fight equally elusive opponents. An analysis of all these additional cases would lead to stronger conclusions about what factors correlate with successful swarming.

ACKNOWLEDGMENTS

I am grateful to several people for their advice and help on this monograph. John Arquilla and David Ronfeldt deserve credit for inspiring and guiding me, as well as for patiently reading over various drafts and contributing their expert opinion. Walter Nelson also read and critiqued an early version, which led to the inclusion of at least two of the historical cases. David Persselin and Russ Glenn read early versions. I owe them a substantial intellectual debt for their expert contribution and critical analysis. Jeff Marquis was indispensable in helping to refine the analysis about military swarming in general. Thomas McNaugher and Michele Zanini deserve thanks for their insightful comments. Paul Davis and Randy Steeb deserve special thanks for their patient and constructive review of the final draft. Finally, I thank my editor, Marian Branch, for her constructive comments and meticulous improvement of the final draft.

Needless to say, any error, omission, or misuse of history is my responsibility alone.

ABBREVIATIONS AND ACRONYMS

AAN	Army After Next
ACTD	Advanced Concept Technology Demonstration
AFQT	Armed Forces Qualification Test
AI	Artificial Intelligence
APC	Armored Personnel Carrier
ASB	Army Science Board
ASDIC	Anti-Submarine Detection and Investigation Committee
ATACMS	Army Tactical Missile System
BAT	Brilliant Anti-Tank
COIN	Counterinsurgency Operations
CS	Combat Support
CSS	Combat Service Support
CTC	Combat Training Center
C2	Command and Control
C3	Command, Control, and Communications
C3I	Command, Control, Communications, and Intelligence
C4ISR	Command, Control, Communications, Computers, Intelligence, Surveillance and Reconnaissance
DARPA	Defense Advanced Research Projects Agency
DCSOPS	Deputy Chief of Staff for Operations and Plans
DIVARTY	Division Artillery
DoD	Department of Defense
EXFOR	Experimental Force
FCS	Future Combat System
FLIR	Forward-Looking Infrared Radar

GPS	Global Positioning System
HEMTT	Heavy Expanded Mobility Tactical Truck
HET	Heavy Equipment Transporter
HMMWV	High Mobility Multi-Purpose Wheeled Vehicle
IFV	Infantry Fighting Vehicle
ISR	Intelligence, Surveillance, and Reconnaissance
JRTC	Joint Readiness Training Facility
JSTARS	Joint Surveillance Target Attack Radar System
LSV	Light Strike Vehicle
MCAN	Marine Corps After Next
MLRS	Multiple Launch Rocket System
MOS	Military Occupational Specialty
MOUT	Military Operations on Urbanized Terrain
MRC	Major Regional Contingency
MTW	Major Theater War
NCO	Noncommissioned Officer
NTC	National Training Center
OPFOR	Opposing Force
OSD	Office of the Secretary of Defense
PSYOPS	Psychological Operations
QDR	Quadrennial Defense Review
RFPI	Rapid Force Projection Initiative
RMA	Revolution in Military Affairs
ROK	Republic of Korea
SNA	Somali National Army
SOF	Special Operations Forces
TDA	Table of Distribution and Allowances
TOE	Table of Organization and Equipment
TRADOC	Training and Doctrine Command
WWI	World War I
WWII	World War II
WMD	Weapons of Mass Destruction

Chapter One

INTRODUCTION

Much of the current discussion about future doctrine for U.S. ground forces involves concepts such as dispersed operations, networking, and greater autonomy for small units than has been customary. One important part of that doctrinal discussion relates to the feasibility and utility of "swarm tactics," tactics that would have small distributed units and maneuverable fires converge rapidly on particular targets. To help inform the debate over the potential relevance of swarming to U.S. military doctrine, this monograph analyzes swarming examples throughout military history. A close reading of those examples might reveal historical patterns and lessons that remain important today.

The research described here was motivated by earlier RAND work on the implications of the information revolution, the advantages that revolution confers on network-based organizations, and the potential value of swarming as a key method of warfare at both the tactical and operational levels.[1] Closely related ideas are being vigorously

---

[1]Two RAND authors, John Arquilla and David Ronfeldt, propose that the information revolution favors the rise of network-based organizations and that swarming will be the major mode of conflict in the future. They propose that the U.S. Army's current AirLand Battle doctrine may need to evolve to a doctrine based on swarming. Their swarming proposal, named "BattleSwarm," is still not completely formulated, but does suggest that smaller and more-maneuverable tactical units be deployed in dispersed networks and trained to use swarming as an operational concept. See John Arquilla and David Ronfeldt, *In Athena's Camp: Preparing for Conflict in the Information Age,* Santa Monica, CA: RAND, MR-880-OSD/RC, 1997. The rise of network-based organizations is discussed in two other Arquilla and Ronfeldt pieces: *The Advent of Netwar,* Santa Monica, CA: RAND, MR-789-OSD, 1996, and "Cyberwar Is Coming!" *Comparative Strategy,* Vol. 12, No. 2, Summer 1993.

pursued by the Marines and certain elements of the Army.[2] If these ideas prove out, it might mean, for example, that the current hierarchical organization of Army field units should be replaced—at least in part—with a hybrid network-hierarchical organization. Divisions and corps would be replaced by smaller maneuver units.

SOME DEFINITIONS

A definition of swarming is necessary before the proper historical examples can be selected. For the purposes of this monograph, a *swarming case* is any historical example in which the scheme of maneuver involves the convergent attack of five (or more) semi-autonomous (or autonomous) units on a targeted force in some particular place.[3] "Convergent" implies an attack from most of the points on the compass.

Admittedly, the phrase "convergent attack" could be stretched to include every case in history in which an army or unit ended up surrounded by the enemy and attacked from all sides during the course of a battle. Encircling and surrounding an enemy has always been a desirable goal: It cuts off the enemy's supply lines and destroys his morale by cutting off any possible retreat. The distinction is that *swarming* implies a convergent attack by many units as the primary maneuver from the start of the battle or campaign, not the convergent attacks that result as a matter of course when some unit becomes isolated and encircled because of some other maneuver.

---

[2]Including the Army's Army XXI and Army After Next (AAN) programs and the U.S. Army War College.

[3]The scheme of maneuver describes how arrayed forces will accomplish the commander's intent. It is the central expression of the commander's concept for operations and governs the design of supporting plans or annexes. Planners develop a scheme of maneuver by refining the initial array of forces, using graphic control measures (i.e., military symbols such as unit icons, phase lines, avenues of attack, etc., usually drawn on acetate and placed over maps) to coordinate the operation and to show the relationship of friendly forces to one another, the enemy, and the terrain. Digitized units in Force XXI will do all this development on a computer screen and avoid the paper and plastic products. See U.S. Department of Defense, *Staff Organization and Operations*, Washington, DC: Department of the Army, FM 101-5, May 1997.

For example, the German *Blitzkrieg* campaigns of World War II (WWII) were not swarming operations according to our definition. Mobile armored warfare was characterized by rapid encirclements, which led, in turn, to convergent attacks on isolated pockets of enemy troops. However, the initial attack and maneuver of the *Wehrmacht* were not convergent. The Germans usually had to concentrate mass before attempting to penetrate opposing lines; after a breakthrough, panzer units usually tried double envelopments or pincer movements.

This particular definition of swarming is useful because it allows the collection of as much empirical data as possible without including every siege and encirclement battle in history.[4] As the analysis compares and contrasts various historical cases, a more sophisticated concept and definition of swarming may emerge. Only by starting with a loose definition of swarming will the analysis proceed to a more informative stage.

It is important to differentiate between swarming tactics and conventional tactics that involve only frontal attacks with one or more flank attacks. For example, single envelopments occur when one army makes a frontal attack to pin the enemy while a mobile part of the force attacks one enemy flank.[5] Sometimes a double envelopment is possible, whereby the enemy front and both flanks are attacked simultaneously (see Figure 1.1).

These traditional set-piece battles are much different from the swarming examples examined in this monograph.

Swarm cases can be broken down into four general categories, based on whether the swarming army begins from a dispersed or a massed position, and whether swarming occurs at the tactical or operational

---

[4]Sieges upon castles, fortifications, and cities can be thought of as convergent attacks, especially because these types of attacks usually succeeded only when the defender was completely surrounded and cut off from all supply. Breaching a defensive perimeter often required the besieger to attack from all sides to distract the defender from the main assault. However, a siege involves little maneuver, except to simply ring the objective with a fortified camp.

[5]Alexander the Great usually tried this using his phalanx as the pinning force, or "anvil," and his flanking cavalry as the "hammer."

RAND *MR1100-11*

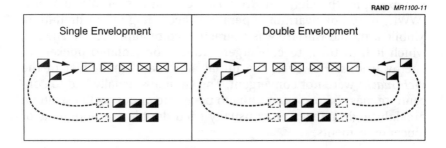

Figure 1.1—Single and Double Envelopments

level.[6] The first approach may be called "Massed Swarm," whereby a swarmer army begins as a single massed body, then disassembles and conducts a convergent attack to swarm the enemy from many directions. Most historical examples are tactical Massed Swarm cases, such as the horse-archer cases. The second approach may be called "Dispersed Swarm," whereby the swarmer army is initially dispersed, then converges on the battlefield and attacks without ever forming a single massed army (see Figure 1.2). The Dispersed Swarm maneuver is more relevant for a network-based organization operating over a dispersed area.

Either of these approaches can be executed at the operational or tactical level (although Dispersed Swarm cases at the operational level have rarely occurred). The four possibilities are pictured in Figure 1.3.

Most historical examples of swarming are tactical cases because of their primitive command, control, and communication (C3) technologies. The communication needs of a tactical swarmer are minor,

---

[6]It might be useful to provide some terminology. War is conducted on three levels. The highest level, strategy, is concerned with delivering the highest possible number of troops to a battle site and denying the enemy the ability to do the same. Tactics are employed at the lowest level of war—the actual battlefield; they are the crucial moves two armies make when close contact has been established. Operational art is the linkage between strategy and tactics; it is the campaign maneuvering required to either seek or avoid battle. Operational-level maneuvers occur at a larger scale than do tactical maneuvers, both in time and distance.

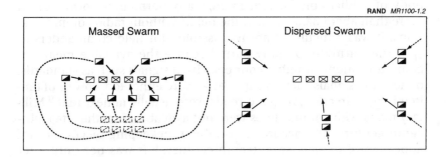

Figure 1.2—Two Types of Swarming Maneuver

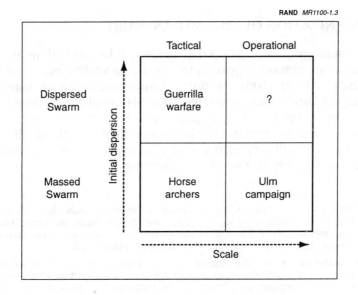

Figure 1.3—The Four Categories of Swarm Cases

assuming all parties on a battlefield can see and hear where the fight is. Operational swarming is much more difficult, because widely separated units must be able to communicate with each other if they are to arrive at the battlefield at the same time from different directions. Before about 1800 A.D., the technological limitations of command prevented army commanders from controlling more than one body of soldiers or, for that matter, any permanent formation of more than about 3,000 men in the field.[7] Without radio communications, it was difficult—if not impossible—for field commanders to split their army into many parts because the available means of communication—whether couriers, visual signals such as standards, or acoustic signals such as trumpets—were either too slow or of limited range.[8] In addition, good roads were usually nonexistent.[9] Military maps with contour lines were not available until the late eighteenth century, and accurate, portable timekeeping pieces did not appear until the late seventeenth century.[10] It was not until after 1800 that these technological building blocks were in place, ready to be exploited by a commander with the genius to recognize them: Napoleon.

ORGANIZATION OF THIS MONOGRAPH

This monograph is organized as follows. It begins with an explanation of the methodology used to select and analyze historical cases (Chapter Two). Next, it notes important research questions (Chapter Three); then generates some historical conclusions from a systematic, brief review of each historical case (Chapter Four). Chapter Five incorporates the lessons of the past with ongoing work by the U.S. military into a discussion of a possible swarming doctrine. Chapter Six provides conclusions.

---

[7]As Martin Van Creveld explains in *Command in War* (Cambridge, MA: Harvard University Press, 1985), the term *formation* is used here to mean any body of men who are effectively controlled. Three thousand men is the densest mass of humanity that can physically see and follow a standard or flag on the battlefield.

[8]The notable exception is the Mongols, who were able to effectively use a combination of "arrow riders" and a mission-order system of command (in which small-unit commanders were granted the freedom to deal with the local tactical situation on the spot while following the overall commander's intent) to assemble their columns at the right place and time. By the second half of the nineteenth century, the combination of the telegraph and railroad also provided some capability to strategically assemble armies in a theater of operations.

[9] The single exception is Roman roads.

[10] Van Creveld, 1985, p. 26.

Chapter Two

METHODOLOGY FOR HISTORICAL ANALYSIS

Military historians cannot manipulate independent variables and conduct randomized experiments to establish causal relationships, as does a physical scientist. History presents observational data, making it impossible for the historian to state that "x causes y." Our goal is to document a relationship or discover an association between two or more variables in the targeted set of cases, without establishing causality.[1]

Usually a victory or defeat in a particular battle depends on a host of factors, and many of those factors are often interrelated. For example, can the historian use case studies in which a nonswarming army defeats a swarming army and attribute that victory to any one variable such as concentration of mass? It is very difficult to know what other variables such as command, logistics, morale, or even blind luck play in determining the outcome. When Crassus's Roman legions were cut to pieces by the swarming Parthians in 53 B.C., was the Roman defeat due to incompetent leadership or the asymmetric balance between highly mobile horse archers and slow-moving infantrymen? Any single variable—such as technology—cannot be viewed in isolation.

---

[1]Research designs that rely on observational data are called *quasi-experimental designs*. Since quasi-experimental designs cannot randomly assign subjects to levels of the independent variable, causal inferences about the effects of independent variables (such as swarming) on dependent variables (such as the battle outcome) are difficult to make. As with all observational datasets, the threat of selection or omitted-variable bias exists. However, designs based on observational data are useful when the goal of the researcher may be simply to explore the distribution of a certain variable in some population of interest. In this case, considerations about internal validity have no relevance.

It follows that, in studying historical cases, it is often helpful to combine observation with an evolving theory—to identify the likely key factors and relationships among factors early in order to be better able to "see" and interpret historical cases with some sophistication.[2] The purpose is to evolve good, structured insights, even if rigorous and precise conclusions cannot be drawn from the historical cases alone.

In this work, it became apparent early on that it would be important to pay attention to the two sides' relative mobility, situational awareness, quality of training and leadership, and weapon characteristics. Other key factors emerged, such as the army's willingness to take losses and the objective of the conflict.

RESEARCH QUESTIONS

The goal of this research is to answer the following research questions:

Historical Pattern-Matching

- When did swarming work and when did it fail?
- Are there any apparently dominant factors?
- How do swarmers fare against nonswarmers?
- Does swarming have particular advantages for offense rather than defense, or does it apply to both?
- How does swarming success vary according to terrain?
- How did swarmers satisfy their logistics demands?
- How have technological limitations on command, control, and communications affected swarming?

---

[2]Qualitative tools can serve a useful purpose. It is possible to draw inferences, eliminate hypotheses, pattern-match, and make analytic generalizations. Because historical data are observational, the historian cannot conclude that the dominant factors that emerge from pattern-matching play a causal role in swarmer battlefield success. However, it is a plausible interpretation not contradicted by the results.

Doctrine

- Can a swarming doctrine be used across the military conflict spectrum, from low-intensity to high-intensity? Or are certain environments and missions more conducive to swarming?
- How does the swarmer avoid defeat in detail?
- What are proven countermeasures to swarming?
- What is the best organizational design for swarming?

   - Should units be combined-arms units?
   - Should swarm units have organic Combat Service Support (CSS) capability?
   - What mix of direct- and indirect-fire capability should swarm units have?

The answers to the historical pattern-matching questions should shed some insight into the doctrinal questions of the future.

SELECTION OF SWARMING EXAMPLES

Examples of swarming can be found throughout military history, from the numerous horse-archer battles in ancient times to the urban street skirmishes in Mogadishu, Somalia, in 1993. Swarming has been employed at the tactical and operational levels, both defensively and offensively, in cities, deserts, jungles, and oceans, by conventional and unconventional forces. This analysis does not look at every swarming example in military history but uses a cross section of cases, each case unique in terms of the force structure of opposing armies and the type of swarming exercised. The goal is to cover what can be described as the prototypical cases, not to exhaustively describe every case that has ever occurred. For example, if the many obscure battles between the Roman legions and the Parthian horse archers usually resulted in a Roman defeat, perhaps the selection of a single battle in 53 B.C. can serve as our "unsupported infantry versus swarming horse archer" prototype.

Several important variables—including weapons, training, terrain, and mission—do not remain constant across the cases. What does remain constant is that one of the belligerents used swarming while the other did not. The ten general cases in this study are listed below in Table 2.1. One case is from a peace operation, two could be classified as "guerrilla actions," and one is a naval case. In another case, swarming was employed both at the tactical and operational levels of war; in yet another, swarming occurred at the operational level, but conventional tactics were employed at the tactical level. Various horse-archer cases against different types of opponents are covered, as is a case of a light infantry swarmer versus a light infantry nonswarmer.

Lessons based on ancient military units like the horse archer are relevant to the evaluation of modern, mechanized-war doctrine because certain principles of military science remain constant no matter what technologies prevailed at the time. For example, the concepts of "standoff firepower" may equate to the composite bow in one age and to the Army Tactical Missile System (ATACMS)—delivered Brilliant Anti-Tank (BAT) submunition in another. Also, because swarming may seem to violate some principles of war, performing a survey of swarming examples helps us understand the conditions under which we are justified in violating (or reinterpreting) those principles.[3]

---

[3]The principles of war art are offensive, mass, economy of force, maneuver, unity of command, security, surprise, and simplicity, as defined in the 1993 FM 100-5, *Operations* (U.S. Department of Defense, Washington, DC: Department of the Army).

Table 2.1

Historical Swarming Cases

Case Study, Specific Battle or Period of Sample Conflict, and Time (assume A.D. unless otherwise noted)	Terrain	Swarmer Mission: Defensive or Offensive?	Swarmer Description	Nonswarmer Description	Uniqueness of Example
Scythians vs. Macedonians, Central Asian campaign, 329–327 B.C.	Steppe, desert	Both	Bow cavalry	Heavy infantry phalanx supported by heavy cavalry	Horse archer against Macedonian phalanx with supporting light cavalry
Parthians vs. Romans, Battle of Carrhae, 53 B.C.	Steppe, desert	Both	Bow cavalry	Heavy infantry in legions	Horse archer against unsupported legions
Seljuk Turks vs. Byzantines, Battle of Manzikert, 1071	Open, rolling	Both	Bow cavalry	Bow cavalry, cataphracts, bow infantry	Horse archers against combined-arms opponent
Turks vs. Crusaders, Battle of Dorylaeum, 1097	Desert	Both	Bow cavalry	Heavy cavalry	Horse archers against heavy cavalry and supporting light infantry
Mongols vs. Eastern Europeans, Battle of Liegnitz, 1241	Steppe, plains	Offensive	Light and heavy cavalry	Heavy cavalry and infantry	Tactical and operational swarming
Woodland Indians vs. U.S. Army, St. Clair's Defeat, 1791	Woods	Offensive	Tribal warriors (light infantry)	Light infantry, some field artillery	Swarming light infantry versus light infantry
Napoleonic Corps vs. Austrians, Ulm Campaign, 1805	Woods, mountains, steppe	Both	The tactical unit was combined arms (musket infantry, cavalry, horse artillery); the operational unit was the semi-autonomous corps	Combined arms (musket infantry, cavalry, horse artillery)	"Operational" swarming combined with conventional tactics
Boers vs. British, Battle of Majuba Hill, 1881	Rolling grasslands	Defensive	Dismounted cavalry	Infantry	Guerrilla warfare with swarming-like tactics
German U-boats vs. British convoys, 1939–1945	Naval	Offensive	"Wolfpacks" of U-boats	Convoys of merchant ships guarded by destroyer teams	Naval example
Somalis vs. U.S. Commandos, Mogadishu, October 3–4, 1993	Urban	Both	Tribal militia (light infantry)	Light infantry, light vehicles, helicopter gunships	Peacemaking operation

NOTE: Cataphracts are heavy cavalry armed with lance, bow, and sword.

Chapter Three

HISTORICAL CASES

PRE-MODERN SWARMING: HORSE-ARCHER CASES

Many examples of military swarming at the tactical level come from the ancient world and the Middle Ages. The most common swarmer in history has been the horse archer, which was introduced into warfare by the nomadic barbarians of Central Asia. Swarmer-versus-nonswarmer battles usually involved light cavalry armies of nomadic people fighting infantry armies from more-settled agricultural communities.[1] The Eurasian steppe produced most of the well-known mounted archers, including the Scythians, Parthians, Huns, Avars, Bulgars, Magyars, Turks, Mongols, and Cossacks.

The firepower and mobility advantages of the steppe warrior were not surpassed until the invention of gunpowder. Whether their opponent was Persian, Macedonian, Roman, Frank, or Arab, mounted archers usually fared well. Unfortunately, many of the ancient examples of swarming offer little detail because of the remoteness of the events and the lack of accurate and complete accounts. There are few ancient or medieval historical sources on the history of warfare between swarmers, because most swarmer armies were nomadic.[2] Often, only a brief description of the conflict is available.

---

[1]Very few military systems used combined-arms forces, so it was not uncommon for one type of single-arm force to meet another very different single-arm force in battle.

[2]Historical matchups between swarming armies (where both sides relied on horse archers) would include the Mongolian campaign in Khwarezm (1219–1221) and the Mongol incursions into the Middle East later in the thirteenth century. The most famous encounter was the Battle of Ayn Jalut in 1260, in which an Egyptian army led by Mamluks halted the Mongols. See Erik Hildinger, *Warriors of the Steppe: A Military History of Central Asia, 500 B.C. to 1700 A.D.,* New York: Sarpedon, 1997, pp. 163–167.

Central Asian Operations of Alexander (329–327 B.C.)

Alexander the Great was one of the first Western military commanders to encounter an enemy who used swarming tactics. The Scythians, a nomadic people who generally fought with horse archers and used swarming tactics, turned out to be the first army to defeat the Macedonian phalanx after it crossed the Hellespont. However, Alexander improvised new tactics to counter the swarming tactics of the Scythian horse archers and eventually defeated them.[3]

After Alexander successfully defeated the Persian Emperor Darius at the Battle of Gaugamela, he turned his attention to securing the northeastern border of the old Persian empire, especially in the two *satrapies* [provinces] of Bactria and Sogdiana (Figure 3.1), where a revolt had erupted under the leadership of Spitamenes.[4]

While Alexander was building a new fort called Alexandria Eschate on the border near the Jaxartes River (in modern-day Uzbekistan), Asiatic Scythians living on the north side of the river appeared and began to taunt and insult Alexander and his fellow Macedonians.[5] With bone splinters still working their way out of his leg (from a wound picked up in a previous battle), Alexander was in a foul mood. He decided to cross the river and attack the Scythians.

The Scythians used what were known as "Parthian tactics," taking advantage of their greater mobility to circle around their enemies and cause their attrition using long-range arrow fire.[6] Encirclement

---

[3]In *The Generalship of Alexander the Great* (Piscataway, NJ: Rutgers University Press, 1960), J.F.C. Fuller offers an excellent analysis of how Alexander improvised his tactics to defeat the Scythians. In his chapter "Alexander's Small Wars," Fuller extrapolates from the classical descriptions (by Arrian of Nicomedia and Quintius Curtius Rufus) and details the logical sequence of tactics shown in Figure 3.3.

[4]It would take Alexander two years of guerrilla fighting to subdue these two regions. Largely because of the inhospitable terrain, Alexander adjusted his Macedonian army force structure to include more cavalry and light troops.

[5]The Scythians were also the Massagetae, a nomadic people who inhabited the steppe beyond the Jaxartes River. See Fuller, 1960, p. 118.

[6]All of the horse archers looked at in this study used some variant of the recurved composite bow, made of sinew and horn to withstand tension and compression. Composite bows were superior to the Western "self" bows made of a single straight stave of wood. Given equal draw weights, the composite bow will shoot an arrow faster and further than will a self bow. Composite, recurved bows are also shorter and better for men on horseback. For an excellent discussion of this topic, see Hildinger, 1997, pp. 20–31.

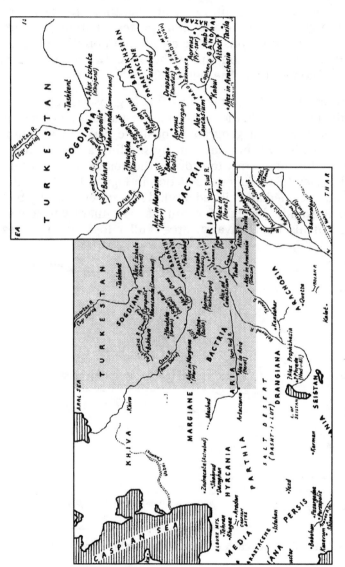

SOURCE: Adapted from J.F.C. Fuller, *The Generalship of Alexander the Great*, Piscataway, NJ: Rutgers University Press, 1960 (reprinted by DaCapo Press, Inc., New York, 1989). p. 110. Copyright© by John Frederick Charles Fuller. Reprinted with permission.

Figure 3.1—Map of Bactria and Sogdiana

maximized the number of targets available to the mounted archers. The general motion of the swarming mass was most likely a slow rotation, which resulted naturally from the individual motion of mounted archers as they continually attacked and retreated (or "pulsed"). Individual riders made short pulses, charging forward from their encircling positions to fire arrows both on the approach and over their shoulders on the withdrawal (where the term "Parthian shot" comes from). See Figure 3.2.

Alexander realized that the best way to come to grips with the more-mobile Scythians was to pin the swarmer against an obstacle, such as a river or a fort. Since a geographic obstacle was not at hand, Alexander used his own men as bait by sending a cavalry force forward before his main army to provoke the hostile horse archers into attacking (see J.F.C. Fuller's depiction in Figure 3.3). Once the Scythians had swarmed and circled around Alexander's cavalry bait as expected, Alexander brought forward his light infantry to screen the advance of his main cavalry force. Fuller logically assumes

RAND *MR1100-3.2*

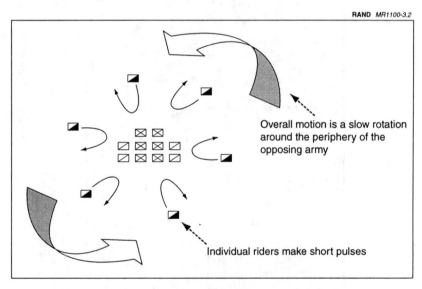

Figure 3.2—The Tactical Motion of Horse Archer Swarming

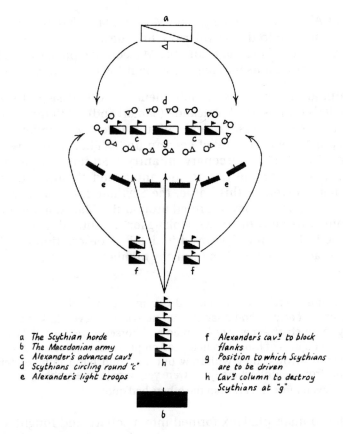

a The Scythian horde
b The Macedonian army
c Alexander's advanced cav!
d Scythians circling round "c"
e Alexander's light troops

f Alexander's cav! to block
   flanks
g Position to which Scythians
   are to be driven
h Cav! column to destroy
   Scythians at "g"

SOURCE: J.F.C. Fuller, *The Generalship of Alexander the Great*, Piscataway, NJ: Rutgers University Press, 1960 (reprinted by DaCapo Press, Inc., New York, 1989), p. 240. Copyright© by John Frederick Charles Fuller. Reprinted with permission.

Figure 3.3—Alexander's Anti-Swarm "Bait" Tactic

that the subsequent cavalry charge was aimed at the Scythians trapped between the light infantry and the bait force. Over 1,000 Scythians were killed and 150 captured in this battle, although the main part of the horse archer army escaped.[7] The Scythians sued for peace shortly thereafter.

---

[7]Fuller, 1960, p. 119.

Without Alexander in charge, the Macedonian phalanx was much more vulnerable to the Scythians. While Alexander was winning his battle at Alexandria Eschate, another Macedonian phalanx was being cut to pieces by a horse-archer army about 150 miles away.

The primary Sogdian rebel, Spitamenes, had laid siege to one of Alexander's outposts at Maracanda (Samarkand). With him were 600 Scythian horse archers. To deal with this threat, Alexander had dispatched a Macedonian relief column under Pharnuches, with 860 cavalry and 1,500 mercenary infantry.[8] Spitamenes lured the Macedonians into the desert and ambushed them south of the Polytimetus River. In this battle, the Scythian horse-archer tactics worked quite well. They swarmed around the Macedonian phalanx and bombarded it with arrows, looking for any subordinate units that could be isolated and destroyed in detail. A description of a horse archer attack could be taken from any number of battles fought then or later:

> The [horse archers] surrounded our men and shot such a great number of arrows and quarrels that rain or hail never darkened the sky so much and many of our men and horses were injured. When the first bands of [horse archers] had emptied their quivers and shot all their arrows, they withdrew but a second band immediately came from behind where there were yet more [horse archers]. These fired even more thickly than the others had done...[9]

The Macedonian phalanx formed into a square and fought a rear-guard action, trying to reach a woody glen and prevent the horse archers from circling. But, in their rush to safety, the troops broke their formation and were annihilated.[10]

When Alexander learned of the disaster, he personally led a combined-arms force of infantry, archers, and cavalry on a march of

---

[8]See Fuller, 1960, p. 242; Peter Green, *Alexander of Macedon, 356–323 B.C.: A Historical Biography*, Berkeley: University of California Press, 1991, p. 357.

[9]Philippe Contamine, *War in the Middle Ages*, translated by Michael Jones, New York: Basil Blackwell, 1984, p. 60.

[10]Robert B. Asprey, *The War in the Shadows: The Guerrilla in History*, New York: William Morris and Company, 1994, p. 6.

135 miles in 72 hours to hunt down Spitamenes, but the mounted swarmers easily dispersed out of his reach. At this point, Alexander decided to target the logistics base of the Scythians. He divided his forces into five mobile columns and began establishing a linked system of military outposts, building hill forts throughout the countryside and concentrating villagers into walled towns. This strategy deprived Spitamenes of provisions and horses, and forced him eventually to abandon his elusive tactics. After Spitamenes lost a pitched battle to one of Alexander's lieutenants, his allies decided to betray him. They cut off his head and sent it to Alexander. All resistance collapsed in Bactria and Sogdiana.

The anti-swarm tactics that Alexander used over 23 centuries ago are similar to modern counterinsurgency doctrine. U.S. Army Field Manual (FM) 90-8, *Counterguerrilla Operations*, instructs soldiers to "locate, fix, and engage." Manuals in the 7-series (FMs 7-10, 7-20, 7-30) order soldiers to "find, fix, and finish" the guerrilla.[11] Modern guerrillas avoid decisive engagements with larger forces, just as the ancient horse archers avoided close battle with the Macedonian phalanx.

Swarming requires superior mobility, an advantage that cavalry clearly possesses over infantry. Historians are interested in how infantry managed to remain the dominant arm for so long, despite its lack of mobility. Between the fifth century B.C. and the battle of Adrianople[12] in 378 A.D., infantry—that is, the Macedonian Phalanx

---

[11]The two techniques to engage elusive foes are either to block positions along likely escape routes or to encircle and cut off all ground escape routes and slowly contract the circle. Variations are possible. One or more units in an encirclement can remain stationary while others drive the guerrilla force against them. This "hammer and anvil" technique was used by Republic of Korea (ROK) forces during Operation *Ratkiller* in the Chiri-san mountains in 1951. See Major Kevin Dougherty, "Fixing the Enemy in Guerrilla Warfare," *Infantry*, March–June 1997, p. 33.

[12]Adrianople (378 A.D.) is generally regarded as the turning point for the decline of infantry as the dominant arm and the ascendancy of cavalry. In this battle, the Roman cavalry on both flanks was routed by the opposing Gothic horsemen, leaving the Roman infantry without cavalry support. With Visigoth infantry attacking the Roman front lines on foot and the Gothic cavalry swarming around the legions in the rear and flanks, the battle became a slaughter (this was not a case of swarming, though).

and the Roman Legion—played the decisive role in warfare;[13] however, swarming cavalry armies managed to defeat infantry armies several times during this period.

Parthians Versus Romans at the Battle of Carrhae (53 B.C.)

One of the exceptions to the rule of infantry dominance was the Battle of Carrhae in 53 B.C., in which Parthian horse archers defeated Roman infantry legions.[14] In the campaign of 55–53 B.C., Marcus Crassus led a Roman army of 39,000 into Parthia to fight a cavalry army of unknown size under Surena, near the town of Carrhae in what is modern-day Syria.

The Roman army was made up mostly of legionaries, with 4,000 light troops and 4,000 cavalry. Crassus at first marched his army along the Euphrates River for resupply by boat and to prevent the enemy from encircling the legions. Eventually, however, he was persuaded by an Arab scout to march out into the plains in pursuit of the Parthians. The Romans formed a hollow square and were surrounded by the Parthian cavalry. After some skirmishing, the horse archers swarmed around the besieged infantry and began delivering arrows and spears from standoff range. As Plutarch describes it,

> The Parthians now placing themselves at distances began to shoot from all sides, not aiming at any particular mark (for, indeed, the order of the Romans was so close, that they could not miss if they would), but simply sent their arrows with great force out of strong bent bows, the strokes from which came with extreme violence. The position of the Romans was a very bad one from the first; for if they kept their ranks, they were wounded, and if they tried to charge, they hurt the enemy none the more, and themselves suffered none the less. For the Parthians threw their darts as they fled, an art in which none but the Scythians excel them, and it is, indeed, a

---

[13]For an explanation of why infantry dominated, see C.W.C. Oman, *The Art of War in the Middle Ages: A.D. 378–1515*, revised and edited by John Beeler, Ithaca, NY: Cornell University Press: 1953 (first published 1885).

[14]The Parni were a nomadic Scythian tribe living between the Caspian and Aral Seas. In 247 B.C., they invaded what is now northern Iran and established the Parthian kingdom. They expanded their domination over all of Iran and Mesopotamia at the expense of the Seleucid Empire.

cunning practice, for while they thus fight to make their escape, they avoid the dishonour of a flight.[15]

Once the Romans realized that the Parthians were being resupplied with arrows by camel trains, they knew they could not withstand the missile barrage indefinitely. Crassus sent his son with a picked force of 6,000 legionaries, cavalry, and auxiliary bowmen in an attack designed to pin down the elusive tormentors. The Parthian cavalry feigned retreat, enticing the small column away from the main body; then, cutting it off, they surrounded and annihilated the entire detachment. The harassment of the main body continued until nightfall, when darkness prevented further missile attack. During the night, most of the Romans managed to retreat to the walled town of Carrhae, while others were cut off and annihilated. The next day, the legions continued their retreat toward the relative safety of the nearby hills of Armenia, where it would be difficult for the Parthian cavalry to operate. Surenas caught up with Crassus and offered a parley, which Crassus was forced to accept because his men demanded it. During the parley there was some sort of scuffle and Crassus was killed; after this, the remnants of his army either surrendered or dispersed. About 5,000 eventually returned alive; 10,000 Romans were captured and the rest killed. Legionaries armed with *gladius* [short sword] and javelin were no match for mounted archers.[16]

By the beginning of the fourth century A.D., cavalry made up about 25 percent of the strength of the Roman army and much higher percentages in the Persian and Arabian armies. The rise of cavalry was enabled by the invention of the stirrup and the appearance of new, heavier breeds of horses in Persia and the steppes of Central Asia. In the East, new heavy lancers now complemented the standard light and heavy horse archers that the Parthian, Central Asian, and Chinese peoples had used all along. The lancers forced an enemy to

---

[15]See Plutarch, *Selected Lives from the Lives of the Noble Grecians and Romans, Volume One,* Paul Turner, ed., Fontwell: Centaur Press Limited, 1963, p. 270.

[16]R. E. Dupuy and T. N. Dupuy, *The Encyclopedia of Military History from 3500 B.C. to the Present,* New York: Harper & Row, 1970, p. 117.

remain in close order, making them more vulnerable to horse archers.[17]

The Byzantines and the Battle of Manzikert (1071)

By the sixth century, the Roman legionary was gradually replaced by the cataphract in the Eastern Roman Empire.[18] Cataphracts were heavy cavalrymen who carried the lance, sword, and shield, as well as the bow, effectively combining firepower, mobility, and shock action. Except for the Frankish and Lombard knights, no horsemen in the world could stand against the heavy Byzantine cataphract. Most of the time, the cataphract proved to be an even match against the Asian and Arab horse archer.[19]

The Byzantine military system deserves a closer look, because its combined-arms armies managed to defeat swarming light cavalry forces many times during its 1,000-year history.[20] Using a combination of bow infantry and cataphracts that negated to some extent the standoff capability of horse-archer armies, the Byzantines managed to defend themselves against the attacks of many types of swarmer armies, including Avar, Turk, Bulgar, Slav, and Magyar.

For example, in the tenth century A.D. the Magyars launched numerous raids from the Hungarian steppe into Byzantine territory.[21] The Magyars did not have a standoff-fire capability: Byzantine

---

[17]Dupuy and Dupuy, 1970, p. 137.

[18]Oman, 1953.

[19] Of course, the tactical matchup between military units is just one reason behind Byzantine success. The Byzantines much preferred bribery, diplomacy, and trickery to actual conflict. Byzantine tactics used a flexible approach and organization that provided for a succession of shocks, which is key to victory in a cavalry combat; as many as five different attacks could be made on the enemy before all the momentum of the Byzantine force had been exhausted. They also loved to perform ambushes, including the "Scythian Ambush," a direct copy of the *mangudai* technique of feigned withdrawal. See Oman, 1953, p. 53, and *Maurice's Strategikon: Handbook of Byzantine Military Strategy*, translated by George T. Dennis, Philadelphia: University of Pennsylvania Press, 1984.

[20]The Byzantine army consisted of heavy and light cavalry, as well as heavy and light infantry.

[21]Magyars fought as the Parthians did against Rome. Armed with javelin, scimitar, and bow, Magyars used superior mobility to harass and wear down their opponents until gaps appeared. They would exploit such gaps to cut off and isolate groups. They

foot archers had a longer range than the Magyar horse archers.[22] However, the Byzantines preferred to close with the Magyars rather than exchange missile fire from a distance. Magyar horse archers could not charge the steady infantry of the Byzantines, whose front rank was made up of spearmen carrying long shields that could stop the scimitar-wielding light horsemen. For their part, the Magyar horse archers avoided close combat against heavier opponents, usually settling for long-range harassment to wear down the enemy before coming to grips. Sometimes a decisive result was impossible and the swarming cavalry had to settle for raiding and looting.

The Byzantines studied their various enemies for weaknesses, including the people they called the Scythians, their primary horse archer enemy. *Maurice's Strategikon*, a Byzantine military manual written around 600 A.D., notes that cold weather, rain, and the south wind loosen the bow strings of the horse archer. In the section called *"Dealing with the Scythians, That Is, Avars, Turks, and Others Whose Way of Life Resembles That of the Hunnish People,"* the *Strategikon* notes that these enemies preferred surprise and the cutting off of supplies to direct force. "They prefer battles at long range, ambushes, encircling their adversaries, simulated retreats and sudden returns, and wedge-shaped formations, that is, in scattered groups."[23] They could also be hurt by a shortage of fodder, which they needed for their vast herd of horses. *Strategikon* warns Byzantine commanders to make sure a geographic obstacle such as an unfordable river is at their rear to prevent the swarmers from encircling them.

Despite this record of success, the most disastrous defeat in Byzantine history came at the hands of the Seljuk Turks, a horse archer

---

inhabited the lower Don Basin in the early ninth century, where they were vassals of the Khazar Turks. Driven away from Turkish tribes by eastern pressure, the Magyars migrated to the lower Danube Valley. Eventually, they migrated across the Carpathians into the middle Danube and Theiss valleys to defeat the Slavic and Avar swarmers and establish the Hungarian nation.

[22]According to Erik Hildinger, Byzantine cavalry carried recurved composite bows. See Hildinger, 1997, p. 77.

[23]*Maurice's Strategikon: Handbook of Byzantine Military Strategy*, translated by George T. Dennis, Philadelphia: University of Pennsylvania Press, 1984, p. 117. The other primary source for Byzantine military tactics is Emperor Leo VI's *Tactica*, written around A.D. 900. For a good discussion of its contents, see C.W.C. Oman, *A History of the Art of War in the Middle Ages, Volume One: 378–1278 AD*, London: Greenhill Books, 1998a (first published in 1924), pp. 187–217.

people: the Battle of Manzikert in 1071. Although the Byzantine capital of Constantinople did not fall until 1453, most historians trace the military decline and eventual defeat of the Byzantine Empire to this one defeat.[24] Manzikert led to the loss of rich provinces in Asia Minor, an area that was a source of economic strength and military recruitment. After this battle, the Byzantine defenses were never the same.

Seljuk Turks operating out of Persia had been raiding the eastern provinces of the Byzantine Empire for many years when Emperor Romanus Diogenes decided to do something about it. In 1071, his army of around 30,000 men maneuvered to engage an approximately equal number of Turks near his eastern territory in Armenia.

The battle of Manzikert occurred on excellent horse-archer terrain, open and rolling. It proceeded in the typical swarmer manner, with the Turks hovering about the Byzantine line, shooting arrows but never closing. Byzantine horse archers tried to return arrow fire, but they were too few and suffered heavily.[25] The mounted Turkish archers stayed out of reach, refusing to close with the Byzantine heavy cavalrymen, pouring a constant deluge of arrows into the Byzantine ranks. At the end of the day, Romanus directed his tired army to withdraw, back to camp. The Turks harassed the retiring columns so much that Romanus ordered his army to turn around again and head them off. At this point, the Byzantine reserve line did not follow orders and continued on its way back to camp. Without a rear guard, the Byzantines were quickly encircled by the horse archers.[26] The horse archers folded in their center and swarmed around the flanks of the Byzantine army, pouring in arrows from three directions. When the Byzantine rear guard deserted, the Turks were able to surround the Byzantine main body and turn an orderly withdrawal into a rout.

Manzikert is another example of the *mangudai* pattern—to pretend to retreat, then encircle and ambush your pursuers from all

---

[24]Other factors also contributed, including a continuing decline in training and discipline and the sacking of Constantinople by the Crusaders during the Fourth Crusade.

[25]Oman, 1998a, p. 220.

[26]Oman, 1998a, p. 221.

directions. It has been a favorite tactic of horse archers throughout the ages (see Figure 3.4).

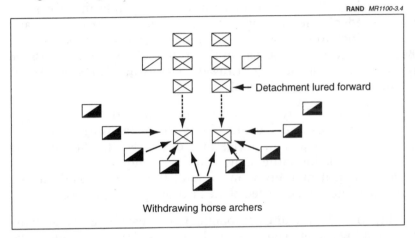

Figure 3.4—The *Mangudai* Technique of Feigned Withdrawal

Again, many reasons can be given for the Byzantine loss to the Turkish horse archers at Manzikert, including the poor leadership of Emperor Romanus Diogenes and some degree of treachery during the battle from one of his reserve commanders. Certainly Byzantine soldiers were not as disciplined as they were during the height of Byzantine power in the sixth and tenth centuries. Byzantine training obviously varied in quality over the course of centuries. But sources indicate that the skillful use of mounted archers and the age-old ploy of the *mangudai* technique by the Turks led directly to the Byzantine defeat.[27]

The First Crusade and the Battle of Dorylaeum (1097)

The Crusades present another unique historical matchup between a swarming horse-archer opponent and the heavily armored knights of Western Europe. Western knights armed with lance and sword

---

[27]Walter Emil Kaegi, Jr., "The Contribution of Archery to the Turkish Conquest of Anatolia," *Speculum*, Vol. 39, No. 1, 1964.

clashed with the Seljuk Turks of Syria and the Holy Land through several Crusades, more often than not suffering at the hands of the horse archers. The First Crusade was the only significant success for the Crusaders—an amazing feat, given their inferior mobility and tactics, which were poorly suited for facing armies of horse archers. Besides a few minor engagements and some siege battles, the Battle of Dorylaeum was the major Crusader victory over the Seljuk Turks. As such, it deserves a closer look.

At the Battle of Dorylaeum (in present-day Turkey) in 1097, two heavy cavalry Crusader detachments caught a swarmer army of Turkish light cavalry in a vise and destroyed it. The following description of the battle shows that even though the Turks had greater tactical mobility man-to-man, their army could still be outmaneuvered and defeated at the tactical-operational level.

During the Dorylaeum campaign, the Crusader army actually marched in separate columns for three days after one of the feudal lords, Bohemond, took his Italo-Norman contingents and separated from it. Bohemond's force probably numbered around 10,000 Crusaders, the majority on foot, along with large numbers of noncombatants. On the evening of June 30, 1097, Bohemond's army made camp in a grassy meadow on the north bank of the River Thymbres, near the ruined town of Dorylaeum.

The next morning, Bohemond's men were attacked by roughly 30,000 Turkish horse archers under the command of Kilij Arslan. The Crusaders had never seen horse-archer tactics before. "The Turks came upon us from all sides, skirmishing, throwing darts and javelins and shooting arrows from an astonishing range."[28] The Franks were shocked to see that every Turk was mounted. The Turks rode around the Crusader camp in loose swarms,[29] killing so many knights with their arrows that Bohemond's army began to retreat toward the banks of the river.

The Turks captured a good portion of the Crusader camp as they swarmed around the Crusaders, cutting off individuals and small

---

[28]Terry L. Gore, "The First Victory of the 1st Crusade: Dorylaeum, 1097 AD," *Military History*, Vol. 15, No. 2, June 1998.

[29]Oman, 1998a, p. 274, and Gore, 1998.

groups. Bohemond ordered his knights to hold their positions. The Turks had the Crusaders virtually surrounded, and they set up relays to keep their archers provided with a constant supply of arrows. Whenever a small detachment of knights charged the Turks, the elusive horse archers would retire just out of reach, sending volley after volley of arrows into the Christians' ranks. Bohemond could only watch as his army died slowly from the "arrows and javelins . . . falling as thick as hail, the savage, piercing shrieks of the enemy, and the diabolical swiftness of their cavalry, constantly darting in to the attack and then away again" (Gore, 1998). The Crusaders were on the verge of defeat.

At this point, some messengers Bohemond had sent earlier to get help finally located the other Crusader detachment and guided them to the battle, where they quickly launched an attack on the Turkish flank and rear. The charging knights caught the Turkish army by surprise and pinned it enough to turn the fight into a melee. Bohemond's tired troops rallied and charged the Turks when they saw their friends charging into the Turkish rear.[30] The battle ended because horse archers were no match for the Western knight in close-quarters battle.[31] Oman notes that total casualties were less than to be expected, because the Turks themselves suffered only during the last 10 minutes of battle before they fled the field.[32]

One of the lessons of the Crusades reinforces the historical pattern: Swarmers must have superior mobility to defeat heavier nonswarmers. If swarmers can be pinned or hemmed in in some manner, they can be defeated.

The Crusaders managed to defeat the swarming army in this battle for two reasons: the failure of the swarmers to keep track of the second Crusader force and the lack of a shock force capable of delivering a knockout blow early on. The victory was due to chance for the most part. Subsequent battles between the Seljuk Turks and

---

[30]Oman, 1998a, p. 277.

[31]Gore, 1998.

[32]Oman, 1998a, p. 277. Gore, 1998, notes that "the Turks found the Western European knight much tougher to kill than the less-armored foot soldier. The knights (who would later be called "iron people" by the Saracens) took numerous missile hits and still fought on."

the Crusaders in later years—such as the Battle of Hattin in 1187—showed that the mounted archer would usually prevail over the Frankish knight.

THE ULTIMATE SWARMERS: THE MONGOLS AND THEIR INVASION OF EUROPE (1237–1241)

The Mongols are the ultimate exemplars of swarming, because they swarmed at both the tactical and operational levels. They defeated swarming and conventional opponents alike. In the early thirteenth century, Genghis Khan defeated all his neighbors and unified Mongolia around the Gobi Desert. Eventually, Mongol conquests stretched from Korea to Germany, the largest continuous land empire ever.

Mongol success can be attributed to many factors, including a decentralized command system that allowed subordinate commanders a great deal of initiative and decisionmaking power.[33] Also, the successful application of the Mongol swarming concept was at least due in part to superior situational awareness, mobility, and standoff fire. Superior mobility came from an army consisting entirely of cavalry, 60 percent of which was light. Standoff fire was enabled by the composite bow.[34]

Mongol light cavalry gathered field intelligence, conducted mop-up operations, pursued the enemy after breakthroughs, and provided

---

[33]Oman attributes Mongol success also to iron discipline (execution was a very common punishment) and the fact that in both Asia and Europe the Mongols faced no principality of great size or strength. See Sir Charles Oman, *A History of the Art of War in the Middle Ages, Volume Two: 1278–1485 AD*, London: Greenhill Books, 1998b (first published in 1924), p. 317.

[34]According to J. Chambers, *The Devil's Horsemen: The Mongol Invasion of Europe* (New York: Atheneum, 1979, p. 57), the Mongol bow compared favorably with its best European counterpart. The English longbow had a pull of 75 pounds and a range of 250 yards; the smaller Mongol recurved composite bow had a pull between 100 and 160 pounds and a range of 350 yards. The Mongols also practiced a technique called the Mongolian thumb lock, whereby an archer used a stone ring on the right thumb to release arrows more suddenly to increase velocity. Hildinger's review of various historical sources and modern experts (1997, pp. 20–31) suggests that the accurate range for shooting the composite bow from horseback is much shorter, between 10 and 80 yards. More inaccurate fire at greater ranges is possible against massed enemies by "shooting in arcade" (shooting at a steep angle of about 45 degrees).

firepower support. Heavy cavalry provided the shock attack option if bow missile fire proved insufficient to destroy the main force. When enemy cohesion was disrupted and gaps appeared during battle, Mongol heavy cavalry armed with 12-foot lances provided the decisive blow.[35]

At the tactical level, the Mongol horse archers used the same methods as their ancient Turkish and Parthian forebears. By fire and maneuver, the more elusive Mongols could remain at a distance, inflicting damage by missile attack. If the Mongols could not encircle the enemy, they tried other tactics, such as the *mangudai* "feigned withdrawal" ruse.

At the operational level, several Mongol divisions, or *toumens*, usually advanced on a broad front in roughly parallel columns (their Hungarian front was 600 miles wide), with a deployed screen of light cavalry to shield Mongol troop movements from enemy observation. Whenever an enemy force was located, it became the objective of all nearby Mongol units. *Toumens* would converge simultaneously on the enemy from multiple directions.[36] The column encountering the enemy's main force would then hold or retire, depending on the situation.[37] Meanwhile, the other *toumens* would continue to advance, approaching the enemy flank or rear. The enemy would naturally fall back to protect its lines of communication and the Mongols would take advantage of any confusion to surround its position.

The Mongol *toumens* avoided defeat in detail by superior mobility and battlefield intelligence. Mongol units were faster because each horseman had several spare mounts to rely upon from the reserve herd of animals that trailed every *toumen* on the march. Riders simply switched mounts repeatedly on the march, as their horses

---

[35]The heavy horsemen also used a scimitar, a battle ax, or a mace.

[36]Separating into *toumens* had two main benefits: It magnified the apparent number of invaders in the panicked eyes of their enemies and it eased the logistics demands, which would be more severe with a concentrated host.

[37]To buy time for other columns to approach, the first column would either pin the enemy if it was strong enough or feign retreat if not.

became exhausted.[38] Despite the vast distances often separating individual *toumens*, the Mongols enjoyed superior situational awareness by using a corps of mounted couriers to relay messages and orders. Tactically, they communicated with signal flags for the most part, but also with horns and flaming arrows. Strategically, Mongol spies were always sent ahead as merchants to the next target region, well before the Mongol *toumens* ever appeared on the horizon.

Mongol success depended on having terrain on which to maneuver.[39] Generally, when the horsemen could swarm around the enemy, they won; when they could be channeled, they lost. [40] Noted historian Sir Charles Oman argued that there were three types of terrain in which horsemen could not fight effectively: marshes, where horses had to follow trails or get stuck; dense woodlands, where horsemen were channeled onto narrow paths; and very mountainous terrain, where movement was restricted to passes. As Oman states, "the Tartar [Mongol] was essentially a conqueror of the steppe and the plainland, and in Europe it was the lands of the steppes and the plains only that he swept over."[41]

In the early thirteenth century, the Mongol empire steadily expanded west, with Russia falling by 1240. After destroying all the Russian *duschies,* the Mongol commander, Batu Khan, set his sights on Hungary. Before he crossed the Carpathians into Hungary, he detached a force under Baidar to watch his northern flank and take care of the Poles. The speed and coordination of the widely dispersed *toumens*

---

[38]Mongol armies could travel 50 or 60 miles a day, several times the distance their European adversaries could travel. See Erik Hildinger, "Mongol Invasion of Europe," *Military History*, June 1997.

[39]Terrain affected logistics as well as mobility. Some authors speculate that the Mongols would never have been able to conquer Germany because they needed open areas with plenty of grass for their herds of horses.

[40]For example, the Mongols learned to avoid mountain passes. The *toumens* with their large herds of backup mounts, could not maneuver easily in the mountains. King Vaclav and his Polish-Czech army defeated a Mongol army in the Silesian passes in 1241.

[41]Oman, 1998b, p. 323.

that knifed into Hungary and Poland bring to mind the armored breakthroughs of WWII.[42]

Two Mongol *toumens* under Baidar met about 25,000 Poles and Germans under Duke Henry II of Silesia at Liegnitz on April 5, 1241. Fighting on fairly open terrain, the Mongols were able to execute one of their favorite ruses, the *mangudai* technique. They managed to lure the European heavy cavalry of Teutonic knights and Templars into a trap by deliberately folding back the Mongol center (which was composed of light horsemen). Once King Henry had committed his other elite heavy cavalry into the attack, the Mongol light horse archers sidestepped the charging knights and enveloped them from three sides, showering the Europeans with a deadly hail of arrows. Smoke bombs added to the confusion. And when the moment was ripe, the Mongols delivered the *coup de grâce* with their heavy cavalry.

After their victory, the Mongols under Baidar rejoined their comrades in Hungary and defeated an even larger European force under King Bela at the Battle of the Sajó, a river in northeast Hungary. Hungary was saved from complete destruction only by the death of the Great Khan Ogotai in faraway Karakorum. Batu Khan led his *toumens* back to take part in the contest for the succession, and the Mongols left Hungary as suddenly as they had entered it.

The Europeans were ill-suited to face the horse archers because of their lack of missile-bearing troops and their poor tactics. Western armies relied upon their heavy cavalry as the main striking force. Its primary purpose was to deliver a decisive charge to break up the enemy formation. Infantrymen played a supporting role, protecting the rear while the knights charged, and finishing off any unhorsed enemy cavalrymen.[43] The Western knights possessed superior armor (plate armor and chain mail), rode stronger horses, and were superbly trained, but they could not close with the faster and lighter Mongol.

---

[42]In fact, both Generals Patton and Rommel admired and studied the principles employed by Subotai, the military commander of the Mongol invasion of Europe in 1240. See Chambers, 1979, p. 66.

[43]Experience and ability varied considerably, from the highly competent detachments of Teutonic knights and Knights Templar from France to the general levy of free peasantry, sometimes armed only with crude farm implements.

INDIAN SWARMING ON THE NORTH AMERICAN FRONTIER: ST. CLAIR'S DEFEAT (1791)

Another historical example of the tactical swarmer is the Native American Indian. In the woodlands of the Ohio Valley territory in the late eighteenth century, Indians possessed superior situational awareness because they knew the lay of the land and used their scouts more effectively than did European-modeled forces. The heavily wooded terrain offered concealment, and the lightly armed Indians were more mobile than the Colonial regular infantry. The Indians used modified swarm tactics to surround the enemy and rush him from all sides. Although they did not have a standoff-fire capability, surprise ambushes based on concealment and superior situational awareness were sufficient to achieve victory.

The worst defeat ever inflicted on a U.S. army by Indians occurred in the Ohio Territory in 1791, at the battle called "St. Clair's Defeat." Nearly 700 American soldiers died in this disaster (three times the number the Sioux would kill 85 years later at Little Big Horn). This example deserves a closer look.[44]

In September 1791, the U.S. commander, Major General Arthur St. Clair, headed north from what is now Cincinnati, Ohio, to establish a string of forts through Indian territory. When his troops were about 50 miles from present-day Ft. Wayne, Indiana, they camped upon some high, defensible ground. A large number of sentries were placed around the bivouac site.

St. Clair received so little intelligence that historians have failed to name the battle in the traditional manner—after the nearest geographic feature—because St. Clair had no idea what river was near his position. As with any historical analysis, many variables affected the outcome. In this case, the Americans were short of horses, their 55-year-old commander had a case of gout, and the attached militia units were poorly disciplined.[45]

---

[44] See Richard Battin, "Early America's Bloodiest Battle," *The Early America Review*, Summer 1996. See also Leroy V. Eid, "American Indian Military Leadership: St. Clair's 1791 Defeat," *The Journal of Military History*, 1993, pp. 71–88.

[45] It is true that a better-trained and better-equipped American Army gained victory over these Indians three years later at the Battle of Fallen Timbers (1794).

But it would be false to conclude that any ragtag army of Indians could have defeated General St. Clair. The complete surprise of the Indian attack on all sides was more to blame for the defeat than mediocre U.S. leadership.

No one knows for sure who was actually leading the Indians. It was either a single, unidentified leader or a council of leaders.[46] In this case, all the Indians appeared to be using the same tactics: charging frontiersmen (who were more apt to break and run than the U.S. regulars), shooting at officers, using the "treeing" technique,[47] and withdrawing and surrounding any U.S. detachments that conducted bayonet charges. As one eyewitness put it, "They could skip out of reach of bayonet and return, as they pleased."

The Indians initially rushed the militia sentries and sent them flying, then a group rushed the main camp. The Indians stayed hidden as they ran through the underbrush and completely surrounded the U.S. camp in a matter of minutes. St. Clair remarked later that he was "attacked in front and rear, and on both flanks at the same instant, and that attack [was] kept up in every part for four hours without intermission."[48] The main weight of the attack was initially in a half-moon shape that overlapped the left flank of the U.S. position, which was the first to collapse.

Individual small-unit leaders followed the same game plan, similar to "the mission-order" of the *Wehrmacht* 150 years later.[49] The psychological effect of the attack broke U.S. resistance. In the end, in the confusion St. Clair luckily managed to punch a hole through the circle of swarmers and escape with a pathetic remnant of his command.[50]

---

[46]The leader was either Little Turtle of the Miamis or Blue Jacket of the Shawnees.

[47]The "treeing" technique was to get down on one knee behind a tree and wait for the appearance of the enemy. The Indian hopped from tree to tree after firing, continuously using one position after another.

[48]Eid, 1993, p. 85.

[49]The Indians were led by different grades of chiefs, some of whom led groups of 50, some 100, etc.

[50]The Indians failed to pursue and finish St. Clair off because they fell to looting the abandoned camp.

ULM: A CASE OF OPERATIONAL-LEVEL SWARMING?

The operational-level maneuver of Napoleon's corps in the 1805 Ulm campaign appears to fall under our broad definition of *swarming*. Several independent and dispersed corps converged simultaneously or "swarmed" from different directions to encircle the Austrian army. This case is unique because it was operational-level swarming only. Ulm was not a battle; it was an operational victory so overwhelming that the issue was never seriously contested in tactical combat. At the tactical level, the French army used an improved version of the standard line-and-column tactics of the day. French forces were made up of infantry, field artillery, and cavalry, and they were applied in tactical offense or defense in the same as were opposing forces.

The heart of Napoleon's system was the *corps d'armée*, the self-contained combined-arms units that could move in a diamond formation of four or five corps (see Figure 3.5).[51] Theoretically, each corps could fight and pin an entire opposing army for at least 24 hours, just enough time for sister corps to converge.[52]

Napoleon sought to assemble rather than concentrate ("assemble" implies a more versatile and flexible stance, one less committed to a particular course of action) major units within marching distance of the intended battlefield on the eve of battle. He then had the flexibility to concentrate mass to whatever degree he chose. His genius lay in being able to balance the requirements of concentration and dispersion to deceive the enemy as to his intentions.[53] He avoided piecemeal destruction using maneuver to both pin the enemy main body and materialize a flanking force on its flank or rear. Napoleon loved to cover the final approach to the enemy with incredibly fast forced marches, "pouncing like a cat."

---

[51]These divisions or corps were first envisioned by Marshal Broglie in the Seven Years War. See David Chandler, *The Campaigns of Napoleon,* New York: The Macmillan Co., 1966, p. 159.

[52]Napoleon liked to scatter sometimes up to a dozen or more major formations, all accessing coordinated roads to converge on the confused opponent. Chandler, 1966, p. 154.

[53]Chandler, 1966, p. 150.

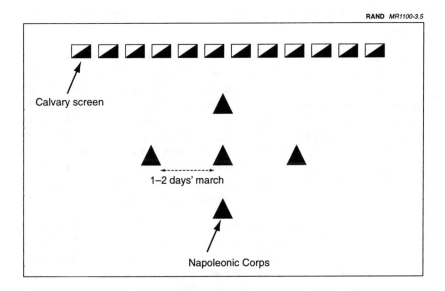

RAND *MR1100-3.5*

Figure 3.5—Napoleon's Diamond Formation

Napoleon once said, "Strategy is the art of making use of time and space." [54] Superior mobility and situational awareness were the keys to using that time and space effectively.[55] Averaging 30 kilometers per day on the march, the French army's speed gave Napoleon the ability to maintain the initiative and stay one step ahead of the enemy. A massive, dense cavalry screen thrown forward would cover

------

[54]When Napoleon said "strategy," he really meant "operational art," a modern term. In those days, the closest equivalent term was strategy.

[55]French units were faster than other armies because they foraged for food on the march and hauled fewer supplies in their siege trains. The social and political changes wrought by the revolution made this possible. The *levee en masse* filled the ranks of the *Grande Armée* with a true cross section of French society. While opposing armies had to rely on mercenaries, conscripts, and general undesirables, the high *esprit de corps* of the French army lowered desertion and granted Napoleon the freedom to spread out his men as much as he wanted. Dispersion allowed them to forage for food and supplies on the move, reducing his need for a logistical tail and increasing his speed. Napoleon's operational art depended on this speed. Cyril Falls, *The Art of War from the Age of Napoleon to the Present Day*, Oxford, England: Oxford University Press, 1961, p. 19.

all movement and limit the enemy's ability to detect Napoleon's penetrating corps.

In 1805, an Austrian army of 72,000 men under the command of Archduke Ferdinand d'Este marched south through southern Germany to the area around Ulm to deny supplies to Napoleon and link up with an approaching Russian army. The effective commander of the Austrian army was the Chief of Staff, General Karl Mack Von Lieberich. General Mack's plan was to act as the "anvil" upon which Napoleon's French army might be destroyed, the 100,000 Russians acting as a "hammer." As Napoleon converged his separate corps toward the Austrians, Mack tried to escape the trap by attacking the French VI Army Corps, mauling a component division in the process.

After the Austrians captured a copy of Napoleon's orders, Mack argued for an immediate move to Regensburg, but Ferdinand delayed him. When the Austrian army finally did move east on October 14, Napoleon was able to stop him at the Battle of Elchingen. Mack had no choice but to hole up in Ulm, where he was operationally surrounded, and he later surrendered with nearly 30,000 men. Figure 3.6 shows the routes of the semiautonomous corps approaching Mack's position from multiple directions.[56]

In contrast, Napoleon's Russian campaign of 1812 illustrates how crippling the loss of operational mobility can be.[57] His final objective was Moscow, which he managed to sack. However, because of logistics problems, he never did manage to get his corps to converge against the opposing Russian army. The Russians adopted a Fabian strategy of scorched-earth withdrawal, avoiding battle when it was advantageous to do so. This strategy deprived the French of even

---

[56]In the broadest strategic sense, this campaign might be viewed by some as a single envelopment or a turning movement, in the way that Mack's major line of communications was severed. The two opinions are not necessarily separate and distinct.

[57]Other factors obviously contributed to Napoleon's defeat in Russia. Napoleon waited too long in Moscow before beginning his retreat, subjecting his troops to the early Russian winter. He chose to fall back to Smolensk, along the ravaged northern route of the original French invasion.

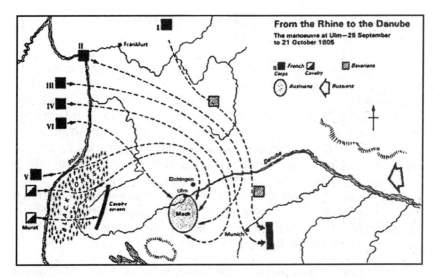

SOURCE: Gunther E. Rothenberg, *The Art of Warfare in the Age of Napoleon,* London: Batsford, 1977, p. 150. Copyright© by Gunther E. Rothenberg. Reprinted with permission.

Figure 3.6—Capitulation of Ulm, October 17, 1805

rudimentary local supplies, forcing them to rely on a burdensome logistical tail. Unlike Western Europe, where Napoleon's operational system could depend on an excellent road network and rich agriculture, the environment in Russia hampered the *Grande Armée's* operational maneuverability (and its swarming ability). The Russians defeated Napoleon in a war of attrition.[58] Napoleon invaded Russia in 1812 with about 450,000 men in his central army group. He returned with 25,000 bedraggled survivors.

---

[58]It is not surprising that the climactic battle of the campaign, Borodino, turned out to be a bloody draw. Over 258,000 men clashed along less than 3 miles. Borodino was a brute-force slugging match with no real maneuver. Both sides used the conventional line-and-attack column as their fundamental tactical deployments.

GUERRILLA OPERATIONS AND SWARMING: MAJUBA HILL (1881)

There may be lessons to learn from looking at swarming examples in the history of guerrilla warfare. In many respects, the tactics used by guerrillas (or insurgents) are characteristic of what a future swarmer might employ. Relying on stealth, surprise, dispersion, and conceal- ment, guerrillas operate without heavy logistical support, move in small groups, and make do without heavy weapons. As to tactics, their favorite offensive approach is the tactical ambush, in which surprise and deception are key. Guerrillas avoid fixed, linear defenses, and they prefer to attack after the opponent has penetrated their defensive area.[59] Most guerrilla examples in history are Dispersed Swarm cases, the most relevant swarming approach for a future network-based organization.

Guerrilla warfare also presents unique aspects that relate directly to a future application of swarming. Partisans and insurgents operate from regional bases situated among a sympathizing population. Insurgents also usually operate in terrain that is inaccessible to heavy conventional units—mountains, forests, swamps, or deserts.[60] These aspects complicate a discussion of U.S. swarming. U.S. forces should never have to depend on indigenous support.

Despite these unique characteristics, guerrilla operations are similar enough to swarming to justify a look at guerrilla warfare in history. One example of guerrilla swarming is the tactics used by the Boers

---

[59]Even the German concept of maneuver warfare recognized the poor ability of linear fixed defense to withstand the concentrated mass of an attacker who chooses his time and place. German maneuver warfare doctrine called for a fluid area defense, charac- terized by a thin forward defensive screen positioned to detect enemy penetrations and a heavy operational reserve ready to establish strongpoints. Strongpoints halted offensive thrusts by the enemy and set it up for counterattacks on the flanks of its penetration corridor. The Germans always sought to create flanks for the enemy so they could envelop it, just as Hannibal did at the battle of Cannae.

[60]Even as late as World War II, the mountains of Greece and Yugoslavia and the for- ests of Poland and Russia were sufficiently inaccessible to afford considerable scope for guerrilla attacks against German-used roads, railroads, and communications. By contrast, no guerrilla movement of any significance was able to arise and maintain itself in any of the technologically advanced Western countries overrun by the *Wehrmacht,* crisscrossed as those countries were by modern roads and telecommuni- cations. Martin Van Creveld, *Technology and War: From 2000 BC to the Present,* New York: The Free Press, 1989, p. 302.

during the Anglo-Boer Wars.[61] Even though the Boers recognized the British as their sovereign in the final peace treaty of the Second Anglo-Boer War, the Boers effectively achieved a military stalemate with their swarm tactics.

The Boers adopted swarming tactics after trying to fight the British in conventional head-to-head fights, learning that the British could bring to bear much greater firepower. They organized into geographical units, commandos, which ranged in size from 300 to 3,000 men. Boer swarming tactics followed the essential formula for guerilla tactical victory: Locate, mass, and attack isolated British detachments, then disperse before any relieving force could arrive.[62]

In general, the Boers usually enjoyed the advantages of mobility, standoff fire, and situational awareness over the British, which allowed them to isolate and attack enemy detachments while avoiding greater concentrations of British Regulars.[63] Most Boers were superb horsemen. They used the Mauser rifle, whose 2,200-yard range was greater than its British counterpart.[64] The loyalty and support of the indigenous population helped the Boers conceal themselves and gather intelligence. The key to their success was to be on the strategic defensive, fighting an enemy on their own territory.

During the major campaign of the First Boer War, Major General George Colley led a small British army into Transvaal territory.[65] After a couple of unsuccessful attacks, Colley decided to seize Majuba Hill, a 2,000-foot-high extinct volcano on the extreme right

---

[61]*Boer* is a Dutch word meaning farmers. Boers were descended from the few hundred immigrants of Dutch, German, and Huguenot origin who settled at the Cape of Good Hope during the late seventeenth century. These frontiersmen lived in scattered family groups through the vast country of the Orange Free State and the Transvaal.

[62]Bevin Alexander, *The Future of Warfare*, New York: W. W. Norton & Company, 1995, p.100.

[63]In the specific case of Majuba Hill analyzed below, the Boers did not have an effective standoff capability.

[64]The Boers had the Mauser in the Second Anglo-Boer War, not in the first. However, even in the First Boer War, the Boers enjoyed a greater effective range because the redcoats were still being trained to fire in volleys in the general direction of the enemy.

[65]This section relies primarily upon Joseph H. Lehmann, *The First Boer War*, London: Jonathan Cape, 1972, and Oliver Ransford, *The Battle of Majuba Hill*, London: John Murray, 1967.

flank of the Boer defense. Majuba was composed of alternate horizontal strata of shale and limestone, deep ravines, masses of rocks and dark mimosa scrub—all of which offered good cover and concealment for attacking troops.

Colley marched out of his main base, Mt. Prospect, with 22 officers and 627 men on the night of February 26, 1881. The British force comprised light infantry from four different regiments, with no machine guns or field artillery. Finding the summit deserted, they moved 354 men into position in and around the summit rim by early morning. The rest dug in at the base of the hill to secure the line of retreat.

The Boers on the laager[66] below were completely surprised. However, once they determined that the British could not fire artillery down upon them, they quickly organized to retake the enfilading position. Joubert, the overall Boer commander, gave the order to retake the hill, but a Boer general had to raise the call for volunteers.[67] The first 50 volunteers raced to the base of the hill, and General Smit led a picket around to the south to contain the British force guarding the British line of advance. Other Boer volunteers galloped up the base of the hill in groups of two or three men. Clusters of Boers looked about to see who would lead them, and two more leaders stepped forward.

The Boers had developed their own tactics for assaulting hills in their earlier wars with the native Africans, zigzagging up the hill from cover to cover while marksmen at the base laid down suppressive fire to cover them.[68] The maneuver elements were led by *burghers*

---

[66]A *laager* is a fortified Boer encampment, usually made by lashing wagons together in a circle.

[67]The Boers were not disciplined, as were European armies. Boers were free to move to any part of a battlefield where they considered themselves most useful. They provided their own rifles and ponies. They wore ordinary dun-brown civilian clothes. They feared close-quarter bayonet fighting and preferred to defend in an extended line, where they could bring to bear their superb marksmanship. Their morale was high, and they were excellent at sizing up and exploiting the tactical nature of terrain.

[68]The Boers were expert shots, having grown up in the Transvaal where the plains were black with game. Even Boer children thought nothing of hitting a running buck from the saddle at 400 yards' range.

intimately familiar with the terrain.[69] One of the two main bodies of troops covered the other with flanking fire while the other moved. As other Boers raced up from the surrounding area, they too joined in the attack and caught up with the assaulting forces. A third party of Boers began moving up the east face.

About 150 Boers maneuvered on all sides while a similar number maintained a fusillade of covering fire from the base (see Figure 3.7).[70] At first, the British were surprised by these bold and aggressive tactics, because the Boers were usually defensive. No one thought the Boers would actually close in for close-quarter battle. The redcoats kept their heads down, but gradually they saw through the smoke that Boers were creeping up right under them. Boer commandos and individual clusters of men advanced slowly and methodically up the slope for about 6 hours.

The forward rim defense under a Lieutenant Hamilton came under attack from the front and rear, and his Scottish Highlander troops starting dropping. British reserves resting in the center were rushed forward. Officers tried to organize firing lines amid all the confusion, noise, and smoke. The British fired in volleys; the Boers fought individually, firing from the shoulder, flopping onto the ground, reloading, and rising up again. Under fire from two sides, the British front line broke ranks and retreated to a new rally point in the middle of the summit plateau. Some Boers melted away from the rear of their main attack and repositioned themselves on the British right flank. Boers also appeared on the left flank along the rim. With bullets flying at them from three sides, the British broke for a final time, with everyone heading straight for the south slope. The Boers pursued them relentlessly, inflicting most British casualties during the headlong flight. All told, British casualties were 96 men killed, 132 wounded, and 56 captured. The Boers suffered one killed and five wounded.

---

[69]A *burgher* is another term for a man in the freelance Boer army.

[70]Other sources describe as many as 450 Boers on the assault. See Ransford, 1967, p. 90.

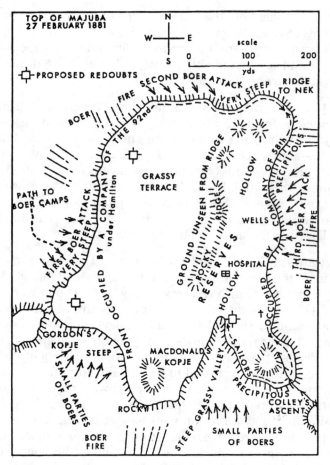

SOURCE: Joseph H. Lehmann, *The First Boer War,* London: Jonathan Cape, 1972,
p. 225.

Figure 3.7—Map of Majuba Hill

Majuba Hill qualifies as a swarming case because semiautonomous
individuals and small units converged on a massed enemy from
nearly all sides. Nonswarmer British light infantry fought from a fixed
defensive position (although they were not dug in) and were deci-
sively defeated. In this case, the Boers were elusive targets, because
they remained concealed as they swarmed on all sides to the top of
the hill. They did not have any standoff capability, but they were

more accurate marksmen. The Boers enjoyed a minor situational-awareness edge because they knew the terrain. When the British came under fire from three sides, their will to fight was broken.

NAVAL SWARMING AND THE BATTLE OF THE ATLANTIC

The historical use of swarming tactics is not limited to land. The German use of U-boat "Wolfpack" tactics during the Battle of the Atlantic (1939–1945) is a naval example of swarming. Packs of five or more U-boats would converge on a convoy of transport ships and their destroyer escorts, independently attacking from multiple directions. British destroyers utilized the ASDIC[71] or sonar to locate U-boats under the surface and counterattacked with depth charges.[72] Whereas in the first half of the war, U-boat Wolfpack tactics proved to be very successful against allied shipping, by 1943 the Allies had perfected a number of technological and tactical countermeasures to Wolfpack swarming. The Germans ultimately failed to win the Battle of the Atlantic. It is important to investigate what caused this reversal.

The Battle of the Atlantic was a battle for superior situational awareness in many respects: Each side was trying to obtain a superior understanding of where the enemy was in relation to its own forces. The vast distances of the Atlantic made this understanding imperative. Strategically, both sides used operation centers that collected and correlated intelligence from all sources worldwide, maintaining great plotting boards. The British tracked German wireless transmissions to try to predict where U-boats were and route convoys clear of them; the Germans did the same in reverse. The U-boat Command in Germany guided U-boats to convoy targets that were located and reported either by electronic espionage, reconnaissance planes, or

---

[71]ASDIC is a British acronym for "Anti-Submarine Detection and Investigation Committee," an early WWII governmental body.

[72]The Allies used three means of detecting U-boats: ASDIC, radar, and HF/DF. The ASDIC or sonar (SOund NavigatiOn Ranging) was a piezoelectric echo ranging device that worked by bouncing a sound pulse off the target. If the echo can be picked up by hydrophones (underwater microphones), a rough bearing and range can be obtained. Radar bounces electromagnetic pulses off objects and notes the origin of the echo. "HF/DF" (pronounced "huff duff"), stands for High Frequency Direction Finding, a device that calculates the direction from which radio messages are sent.

pre-stationed U-boats. The great difficulty for the Germans was finding convoys in time to form a U-boat group in position to attack.

Radio communications allowed the Germans to perfect the tactics of the Wolfpack. U-boats ordered to the area of the reported sighting would spread out in a scouting line across the expected convoy route. The first boat to sight the convoy would begin shadowing it over the edge of the horizon by day, closing at dusk. The U-boat Command located in France would then direct all adjacent boats (within hundreds of miles) to rendezvous with the shadowing U-boat. Once assembled near the convoy, U-boat Wolfpacks preferred to attack simultaneously from multiple directions at night.[73]

Since U-boats could not be detected by ASDIC when they were on the surface and they could outrun all escorts except destroyers, they usually surfaced just before closing with the convoy. After reaching a firing position, most U-boats increased to full speed, fired a salvo of four torpedoes, turned away, fired stern torpedoes if fitted, then retired as rapidly as possible on the surface. After disengaging, U-boats would reload, regain a firing position, and attack again.[74] During the attack, no senior officer was in tactical command.[75] Each U-boat CO attacked as best he could without attempting to coordinate his movements with those of any other boats.[76]

The British Anti-Submarine Warfare Division tried to combat these impulse tactics with various tactical countermeasures. Star shells were used to illuminate the area at night and force U-boats underwater, where they could be detected by destroyers using ASDIC and attacked with depth charges. More escorts were assigned to each

---

[73]In 1940–1941, a typical Wolfpack numbered five to seven U-boats. At first, only one Wolfpack was operational at a time; by August 1942, there were 50 U-boats on patrol and another 20 on passage (out of 140 that were operational), so several Wolfpacks could operate. By February 1943 100 U-boats were at sea. In March 1943, the largest Wolfpack ever (40 U-boats) attacked convoys HX229 and SC122. Vice Admiral Sir Arthur Hezlet, *The Submarine and Sea Power*, New York: Stein and Day, 1967, p. 182.

[74]Hezlet, 1967, p. 167.

[75]The Germans decided that a command boat on the scene was not a good idea, because it could be driven deep and prevented from receiving signals or sending instructions during the battle. Control could be best exercised ashore.

[76] See Peter Padfield, *War Beneath the Sea: Submarine Conflict During World War II*, New York: John Wiley & Sons, Inc., 1995, p. 93.

convoy. Improved radio telephone communication was installed on surface escorts and aircraft.

Tactical situational awareness varied as each side countered the other's detection systems with a series of counter- and counter-countermeasures.[77] Exploiting an early weakness of U-boats—their design to operate on the surface and submerge only for evasion or for rare daylight attacks[78]—the radar proved to be the most important anti-submarine device. Radar could detect German surface attacks at night. Late in the war, the Germans added the *Schnorchel* [snorkel], enabling the U-boat to travel faster underwater. But its speed was still limited.[79] Eventually, Allied aircraft, using radar and depth charges, proved to be a decisive antisubmarine weapon. At first planes did not have the range to cover convoys over the dangerous Middle Atlantic "gap"; ultimately, very long-range aircraft and escort aircraft carriers provided complete air cover across the entire Atlantic Ocean.[80]

U-boats relied on concealment to survive. After 1943, Allied aircraft armed with radar and depth charges seriously constrained the U-boats' ability to remain elusive. Although Allied shipping losses continued to increase until the last year of the war, the Germans were not able to cut the Allied supply line to Europe.

---

[77]The Allies developed the ASDIC, which was partially countered by the Germans when they started using gas bubbles to produce false alarms. To locate surfaced submarines, the British employed aircraft and ship-based radar, along with high-frequency radio direction finders. The Germans responded with search receivers that warned submariners of such surveillance, and later with *Schnorchels*, allowing them to run submerged on their diesels to avoid search radars.

[78]The reason radar was effective was that early in the war, U-boats had to spend most of their time on the surface while traveling. U-boats used a combination of diesel and battery power. Diesel power was the most efficient propulsion (around twice as fast as battery power), but it required the U-boat to surface to take in air for the engines and vent the exhaust. When submerged, the U-boat ran its electric motors on battery power, which made it much slower and limited the time it could remain submerged. Its batteries were recharged when it was running its diesel engines on the surface.

[79]Later in the war—in early 1945—a new Type XXI electric U-boat was finally deployed with a built-in *Schnorchel* capable of staying under water indefinitely, but it was too late to make an impact on the war. The Type XXI could operate underwater at all times (coming up to use its *Schnorchel* once every 4 days), had a new rubber skin, new search receiver, better speed, and new torpedoes.

[80]See Henry Guerlac and Marie Boas, "The Radar War Against the U-Boat," *Military Affairs*, Vol. 14, No. 2, Summer 1950.

The early success of U-boat Wolfpacks illustrates how the advantages of concealment and situational awareness alone were sufficient to overwhelm a convoy's defenses. Once U-boats had converged on the target, coordination in the attack was practically unnecessary. However, the airborne radar seriously undermined the U-boats' elusiveness, forcing them underwater where they lost what little mobility they had. Since the U-boats themselves also served as the primary reconnaissance for U-boat Command, German situational awareness was also undermined.

SWARMING IN PEACE OPERATIONS: BATTLE OF THE BLACK SEA (1993)

The end of the Cold War has seen a dramatic increase in deployments for peace operations, humanitarian assistance, disaster relief, and other small-scale contingencies. Between 1945 and 1989, the Army conducted two peace operations: in the Dominican Republic and in Egypt. Since 1989, it has conducted no less than six such operations (Iraq, Somalia, Haiti, Macedonia, Bosnia, and the Sinai).[81] During Operation Restore Hope in Somalia, U.S. forces fought the most intense infantry firefight since the Vietnam War, against an enemy that used swarm tactics.

Somalia is an important case for the Army and the Marine Corps, because it is the most recent battle in the Military Operations in Urbanized Terrain (MOUT) environment. In an increasingly urbanized world populated by Third World armies using unconventional tactics such as swarming, Somalia is a likely prototype for future peacemaking operations. As such, it makes an excellent case study on swarming.

On the night of October 3, 1993, an assault force of 75 U.S. Rangers and 40 Delta Force commandos fast-roped[82] from 17 helicopters

---

[81]Jennifer Morrison Taw, David Persselin, and Maren Leed, *Meeting Peace Operations' Requirements While Maintaining MTW Readiness*, Santa Monica, CA: RAND, MR-921-A, 1998, p. 5.

[82]*Fast-roping* involves sliding rapidly down very thick nylon ropes hanging from helicopters, usually about 50–100 feet off the ground.

onto a gathering of Habr Gidr clan leaders in the heart of Mogadishu, Somalia.[83] The targets were two top lieutenants of warlord Mohamed Farrah Aideed. The plan was to secure any hostages, and transport them 3 miles back to base on a convoy of 12 vehicles. What was supposed to be a hostage snatch mission turned into an 18-hour firefight over two Blackhawk helicopter crash sites (see Figure 3.8). Eighteen Americans were killed in the fighting.

The dismounted light infantry forces were armed with small arms; the relieving convoys had nothing heavier than HMMWV (High Mobility Multi-Purpose Wheeled Vehicle)–mounted 50-caliber machine guns and automatic grenade launchers. Close air support consisted of Blackhawk and Little Bird (AH-6) gunships. Somalis were armed with assault rifles and rocket propelled grenades.

The Somalis anticipated that after the Rangers fast-roped in they would probably not leave via helicopters (the streets were very narrow). This meant a relief convoy would be necessary, so they immediately began setting up roadblocks all over the city.

The mission proceeded well for the Americans at first. Twenty-four Somali prisoners were quickly seized at the target house. Unfortunately, the mission changed dramatically when a Blackhawk helicopter (Super 6-1) was shot down four blocks east of the target house. Soon after, a second Blackhawk (Super 6-4) piloted by Mike Durant was also shot down about a mile away. An airmobile search-and-rescue force was sent to the Super 6-1 crash site and a light infantry force fast-roped down to secure the wounded crew. Task

---

[83]The picture of Mogadishu and much of this section is drawn from the series of articles published in the *Philadelphia Inquirer* in November and December 1997 by Mark Bowden and from his book, *Blackhawk Down: A Story of Modern War*, New York: Atlantic Monthly Press, 1999.

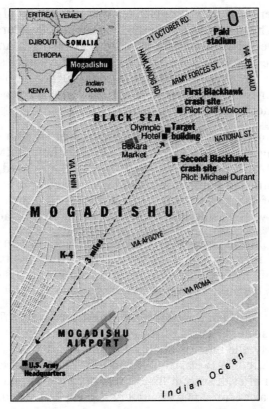

SOURCE: Mark Bowden, *"Blackhawk Down," The Philadelphia Inquirer,* November 2, 1998, http://www.philly.com/packages/somalia/graphics/2nov16asp, and Matt Ericson, *The Philadelphia Inquirer,* staff artist. Reprinted with permission.

Figure 3.8—Map of Mogadishu

Force Ranger was also ordered to move to Super 6-1's crash site and extract the wounded crew.[84] No rescue force was available to secure the second site, which was eventually overrun.[85]

---

[84]Eventually, the Quick Reaction Force of four Pakistani tanks, 28 Malaysian Armored Personnel Carriers (APCs), and elements of the 10th Mountain Division would battle through barricades and ambushes to reach Task Force Ranger at 1:55 a.m. on October 4. See Rick Atkinson, "Night of a Thousand Casualties; Battle Triggered U.S. Decision to Withdraw from Somalia," *Washington Post,* January 31, 1994a, p. A11.

[85] If there was a flaw in the mission planning, it was the lack of a second rescue force. Nobody had taken seriously the prospect of two helicopters going down.

The convoy holding the 24 Somali prisoners was ordered to secure the second crash site, but it never made it. It wandered around, getting chopped to pieces; it eventually aborted the rescue attempt and returned back to base. At one point, after about 45 minutes of meandering, this convoy ended up right back where it started. A second convoy of HMMWVs and three 5-ton flatbed trucks was dispatched from the airport base to attempt a rescue at Durant's downed Blackhawk. But those vehicles were also forced to turn back under heavy fire. Somalis would open fire on any vehicle that crossed an intersection.[86]

For the most part, the commandos followed standard doctrine for city fighting. Using fire and maneuver, teams and squads leapfrogged each other, providing each other fire support in turn. Infantry moved out on foot to cover the convoy from both sides of the street. The main problem was that the convoy kept halting, exposing those vehicles located in the middle of street intersections to concentrated enemy fire.

There was a Somali battle plan of sorts. Aideed's Somali National Army (SNA) militia (between 1,000 and 12,000 men) was organized to defend 18 military sectors throughout Mogadishu. Each sector had a duty officer on alert, connected into a crude radio network.[87] By the time the U.S. assault team had landed, the Somalis were burning tires to summon all militia groups.

The most likely tactical commander of the October 3–4 fight was Colonel Sharif Hassen Giumale, who was familiar with guerrilla insurgency tactics. Giumale's strategy was to fight the Americans by using barrage rocket-propelled grenade (RPG) fire against the support helicopters, ambushes to isolate pockets of Americans, and large numbers of SNA militiamen to swarm the defenders with sheer numbers.

---

[86]Fortunately for the Americans, the ambushes were poorly executed. The correct way to ambush is to let the lead vehicle pass and suck in the whole column, then open fire on the unarmored flatbed trucks in the middle. The Somalis usually opened up on the lead vehicle. They also cared little for fratricide. Because Somalis fired from both sides of the street, they certainly sustained friendly-fire casualties.

[87]Rick Atkinson, "The Raid that Went Wrong; How an Elite U.S. Force Failed in Somalia," *Washington Post*, January 30, 1994b.

Somali tactics were to swarm toward the helicopter crashes or the sound of firefights. Out in the streets, militiamen with megaphones shouted, "*Kasoobaxa guryaha oo iska celsa cadowga!*" ["Come out and defend your homes!"]. Neighborhood militia units, organized to stop looters or fight against other enemy clans, were united in their hatred of the Americans. When the first helicopter crashed, militia units from the surrounding area converged on the crash sites, along with a mob of civilians and looters. Autonomous militia squads blended in with the masses of looters and "civilians," concealing their weapons while they converged on the Americans.

Most of the tribesmen were not experienced fighters. Their tactics were primitive. Generally, gunmen ducked behind cars and buildings and jumped out to spray bullets toward the Rangers. Whenever Americans moved, the Somalis opened up from everywhere. Gunmen popped up in windows, in doorways, and around corners, spraying bursts of automatic fire.

The lightly armed Somali tribesmen who rushed toward the downed Blackhawk helicopters enjoyed two distinct advantages: situational awareness and concealment.[88] They knew where the enemy was, and their approach was concealed.[89] The guerrillas did not need superior mobility. They were on foot but able to keep up with the U.S. convoys, fighting through roadblock after roadblock.[90] Mobility and standoff capability were irrelevant in this case.

With the support of the noncombatants and the intimate knowledge that comes from fighting in their own backyard, clan leaders knew more about what was going on than did the Rangers taking cover in their HMMWVs. Somali women and children acted as sensors, walking right up the street toward the Americans and pointing out their positions for hidden gunmen.

---

[88]They were armed with a mix of Soviet bloc and NATO assault rifles, machine guns, RPG-7s, mines, and demolitions.

[89]The urban terrain limited the effectiveness of close air support.

[90]Gunmen ran along streets parallel to the convoy, keeping up because the two 5-ton trucks and six HMMWVs were stopping and then darting across intersections one at a time. This gave the gunmen time to get to the next street and set up to fire at each vehicle as it came through.

Armed Somali men deliberately used noncombatants, including women and children, for cover and concealment, because they knew the Americans had been issued strict rules of engagement.[91] Rangers were under orders to shoot only at people who pointed weapons at them. Somali soldiers found it easy to blend into gathering onlookers, using noncombatants as cover while they moved toward the crash sites.[92]

U.S. situational awareness was poor. Although officers circling above in command helicopters had access to real-time video during the firefight, the video did not properly communicate the raw terror and desperation of the situation on the ground. Naval reconnaissance aircraft had no direct line of communication with the convoys on the ground.[93] Their attempts to guide the wandering line of vehicles toward the helicopter crash sites failed because of the delay in relaying directions to the ground commander. Pockets of Rangers and "D-boys" [Delta Force soldiers] holed up in adjacent buildings were literally fighting for their lives; oftentimes, they were unaware that friendly units were close by.

From a military viewpoint, the October battle in Mogadishu was a tactical defeat for the Somalis in that the Ranger and Delta commandos were able to complete their mission and extract the hostages. In relative casualties, the mission was also an American military success: Only 18 American soldiers were killed and 73 wounded, while more than 500 Somalis died and at least 1,000 were put in the

---

[91]At one point, a Ranger saw a Somali with a gun prone on the dirt between two kneeling women. He had the barrel of his weapon between the women's legs, and there were four children actually sitting on him. He was completely shielded by noncombatants.

[92]It should be noted that *both* sides may have used noncombatants in Somalia. Somali eyewitnesses have charged that Somali women and children were held as "hostages" by the Americans in four houses along Freedom Road during the firefight, which prevented Giumale from using his 60mm mortars to bombard and destroy the American position around the Super 6-1 site during the night. U.S. officers disputed the notion that Somali mortars would have wiped out Task Force Ranger, because U.S. anti-mortar radar and Little Bird gunships loitering overhead would have destroyed any mortar crew after firing one or two rounds. See Atkinson, 1994a, p. A11.

[93]The Orion pilots were not allowed to communicate directly with the convoy. Their orders were to relay all communications to the Joint Operations Center (JOC) back at the beach. Also, no direct radio communications existed between the Delta Force ground commander and the Ranger ground commander.

hospital—a kill ratio of 27:1.[94] However, from a strategic or political viewpoint, the battle was a swarmer success because the end result was an American withdrawal from Somalia. On November 19, 1993, President Clinton announced the immediate withdrawal of Task Force Ranger and pledged to have all U.S. troops out of Somalia by March 31, 1994. The casualties incurred were simply too high for the U.S. national interests in Somalia.

In this case, the decisive factors that led to a swarmer victory appear to be elusiveness (based on concealment) and superior knowledge of the terrain. Concealment came from the nature of the urban environment, the support of the indigenous population, and the restrictive rules of engagement for U.S. forces. The absence of Somali standoff capability made no difference because the Somalis did not care about casualties.

In the final analysis, the autonomous Somali militia units were able to swarm around the crash sites and the convoys and inflict politically unacceptable losses on a U.S. light infantry force because they were elusive and they enjoyed equal-to-superior situational awareness.[95]

---

[94]Atkinson reported the same number of Americans killed but 84 wounded. He also reported 312 Somali dead and 814 wounded. See Atkinson, 1994a, p. A11.

[95]It is difficult to know for sure what difference an AC-130 gunship or several Bradleys would have made on the outcome. With the presence of noncombatants and the danger of surface-to-air missiles, the gunship may have been of limited value. Bradley Infantry Fighting Vehicles certainly would have provided much greater protection from the RPG and small arms fire than the vulnerable HMMWVs did. The question remains whether the Somalis would have been disciplined and organized enough to swarm RPG fire toward selected Bradley targets.

Chapter Four

HISTORICAL CONCLUSIONS

Across the identified cases, at least three factors appear to play a role in whether or not swarming was successful:

- Elusiveness—either through mobility or concealment

- A longer range of firepower—standoff capability

- Superior situational awareness.[1]

When all three factors were present, swarmers stood a very good chance of winning.[2] Table 4.1 indicates that superior elusiveness and situational awareness appear to be more important than standoff capability.

Certainly one could argue for the inclusion of several other variables. Willingness to take casualties was probably a factor in the Dorylaeum and Mogadishu examples. Training is usually a key variable in most battles. Shooting arrows accurately from horseback is a skill that usually is reserved for those with a nomadic lifestyle, which offers a lifetime of training.

---

[1]Concealment is closely related to superior situational awareness. By definition, *superior situational awareness* involves having more information (unit locations, activity, intent, etc.) about the enemy than he has about you. It is more difficult to conceal your location from the enemy when his situational awareness is superior to yours.

[2]It is worth noting that the Macedonian-Scythian case is exceptional. The swarmer army was defeated despite its having many advantages, because all the advantages in the world matter little against a military genius like Alexander the Great.

Table 4.1

Swarmer Advantages in Ten Specific Battles

Swarmer vs. Nonswarmer	Elusiveness	Standoff Capability	Superior Situational Awareness	Swarmer Strategic Outcome
Scythians vs. Macedonians (Alexandria Eschate)	Yes	Yes	Yes	Loss
Parthians vs. Romans (Carrhae)	Yes	Yes	Yes	Win
Seljuk Turks vs. Byzantines (Manzikert)	Yes	No	No	Win
Turks vs. Crusaders (Dorylaeum)	Yes	Yes	No	Loss
Mongols vs. Europeans (Liegnitz)	Yes	Yes	Yes	Win
Napoleonic Corps vs. Austrian army (Ulm)	Yes	No	Yes	Win
Woodland Indians vs. U.S. Army (St. Clair's Defeat)	Yes	No	Yes	Win
Boers vs. British regulars (Majuba Hill)	Yes	No	Yes	Win
U-boats vs. destroyers (Battle of the Atlantic, 1939–1942)	Yes	No	Yes	Win
U-boats vs. destroyers/aircraft with radar (Battle of the Atlantic, 1942–1945)	No	No	No	Loss
Somalis vs. Rangers (Mogadishu)	Yes	No	Yes	Win

According to this simple pattern-matching, elusiveness and situational awareness appear to be the most important factors in the success of swarming. In Figure 4.1, the intersection of the elusiveness and situational awareness circles is the most crowded region. Standoff capability was important for the horse-archer cases, but less so for the modern infantry cases: The small arms of the Somalis, Boers, and Indians were essentially the same as their opponent's weapons.

Elusiveness allows a swarmer to converge on the enemy in coordination with friendly units when it is advantageous to do so. The historical cases also reinforce the notion that attacks from three or more sides create killing zones, in which both the means and the will to fight are quickly destroyed.

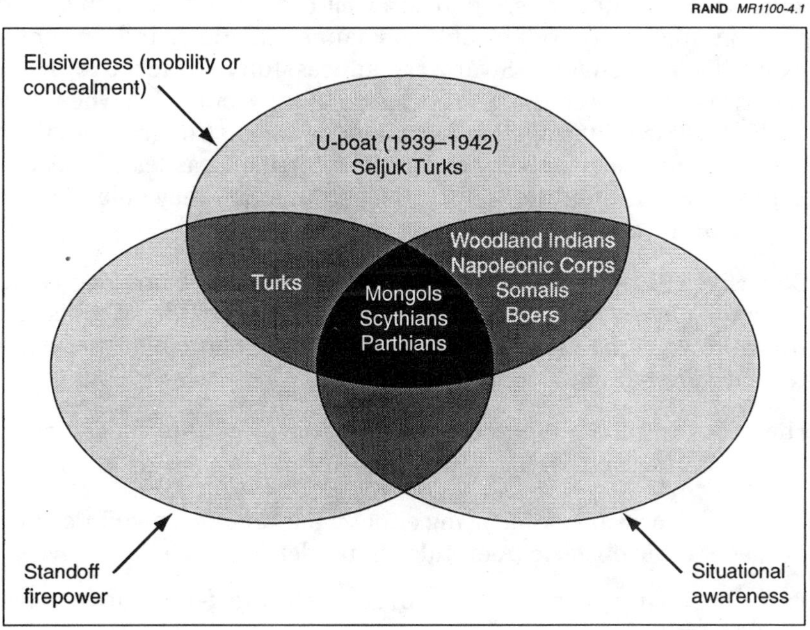

RAND *MR1100-4.1*

Elusiveness (mobility or
concealment)

U-boat (1939–1942)
Seljuk Turks

Woodland Indians
Napoleonic Corps
Turks   Somalis
Mongols   Boers
Scythians
Parthians

Standoff
firepower

Situational
awareness

Figure 4.1—What Advantages Do Swarmers Need?

Some limitations to swarming tactics are apparent. Swarmers were sometimes incapable of a quick knockout blow. The capability for shock action appears to enhance swarmer effectiveness. Most swarmer armies had to wear down their opponents through attrition and standoff fire. The Mongol tactical example is one of the exceptions: Although they usually liked to soften up the target first, their heavy cavalry was capable of delivering the *coup de grâce*.

Terrain appears to have constrained swarming tactics. Swarmers that relied on cavalry for their mobility required terrain with maneuver space (roads, grazing lands, desert, and open plains). Swarmers that relied on concealment for their elusiveness, such as light infantry and U-boats, required concealing terrain such as oceans, forests, and cities.

The historical record of swarming attacks on fixed defenses is a mixed one, so its implications for doctrine are still in question. Using minefields and other obstructions, a well-prepared defense in a fixed position might very well be able to channel an attack and prevent a swarming maneuver.[3] Swarmers successfully attacked a fixed defense when they could remain elusive (Boers), but failed when they could not (Scythians against Alexander's walled strongpoints). Even the Mongols had trouble storming the fortified castles of Eastern Europe, such as after the victory of Liegnitz, when they failed to take Breslau or the castle of Liegnitz.

Defensive swarming must necessarily be porous to some degree, to allow the enemy to penetrate home territory so that local units can swarm toward the invader (for example, Boers, Somalis, and insurgents in general).

The cases have highlighted some successful countermeasures to swarm tactics, such as

- pinning a swarm force using either a part of one's own force or a geographic obstacle (Alexander, Crusaders)

- eliminating the swarm force's standoff-fire advantage (Byzantines)

- eliminating the swarm force's mobility or elusiveness advantage (U-boats)

- securing the countryside by building a linked network of fortifications (Macedonians)

- separating the swarmer from his logistics base (Macedonians).

The effect of swarming on morale is an interesting factor to consider across the observed cases, but it is hard to measure. When a swarming army attacks a defender from all sides, it appears to have an unnerving psychological effect. It is well established that morale wavers when soldiers come under attack from the flanks and/or rear in addition to the front.[4] Soldiers like to know that they have an

---

[3]Probably the most difficult type of swarm maneuver is an attack on a prepared defense in an urban area.

[4]One could foresee a situation in which a swarm attack that completely surrounds an army causes the defenders to fight with even more desperation. If a "hole" is left in the

escape route and that their lines of supply remain open. Because fear is contagious, most battles are won or lost in the minds of the participants, long before the losing side is physically destroyed. Almost all armies have a breaking point in terms of casualties incurred. Coming under a swarm attack appears to lower that breaking point.[5]

In general, command, control, and communications have been primitive in pre–twentieth century swarming cases. Without wireless communication, it is difficult to coordinate many units without keeping them within sight of each other. Interestingly, in all these cases except one—Mogadishu—the swarmer unit was semi-autonomous. See Table 4.2.

Elusiveness was usually based on mobility—the use of horses; in more-modern cases, light infantry used concealment to remain elusive. Rarely did elusiveness stem from both mobility and concealment: Sometimes light infantry were more mobile, sometimes not. Insurgents and Indians were more mobile because they carried less gear and knew the terrain. However, the Somalis were not more mobile. In the U-boat case, concealment was gained at the price of mobility. There is no land example of vehicle-based swarming. Table 4.3 summarizes observations on elusiveness.

Logistics has always been a big challenge for swarmers (as well as for conventional armies). Even when insurgent swarmers relied on the indigenous population and the countryside, they rarely fielded major forces for any sustained campaigning. For the operational-level swarmers such as the Mongols and *La Grande Armée*, a logistics breakthrough was necessary. The Mongol horsemen used immense herds of replacement horses; they were limited to some extent by the availability of good grazing land for their herds. As well, the Mongol soldiers themselves were incredibly hardy individuals. They were

---

circle of attackers, it encourages men with low morale to flee for their lives. The Mongols liked to leave a hole, which usually was set up along an ambush route.

[5]Napoleon once said that "the morale is to the physical as three to one." Morale has always been important in war; however, in this study, it was difficult to look for patterns in morale because of the paucity of historical records.

Table 4.2

Swarmer Command and Control in the Past

Swarmer vs. Nonswarmer	Degree of Autonomy Between Swarmer Units	Amount of Communication Between Swarmer Units	Nature of Communication Between Swarmer Units	Communication Technology
Scythians vs. Macedonians	Semi-autonomous	Little	Tacit	Voice, signals, and standards
Parthians vs. Romans	Semi-autonomous	Little	Tacit	Voice, signals, and standards
Seljuk Turks vs. Byzantines	Semi-autonomous	Little	Tacit	Voice, signals, and standards
Turks vs. Crusaders	Semi-autonomous	Little	Tacit	Voice, signals, and standards
Mongols vs. Eastern Europeans	Semi-autonomous	Moderate	Explicit	Voice, signals, and standards; courier pigeon; human run-ners; smoke
Woodland Indians vs. U.S. Army	Semi-autonomous	Moderate	Explicit	Cavalry runners
Napoleonic Corps vs. Austrians	Semi-autonomous	Little	Tacit	Voice, signals, human runners
Boers vs. British	Semi-autonomous	Little	Tacit	Voice, human runners
U-boats vs. British destroyers	Semi-autonomous	None	Explicit	Radio
Somalis vs. U.S. Commandos	Autonomous	Little	Explicit and tacit	Voice, cell phone

Table 4.3

The Nature of Elusiveness

Swarmer vs. Nonswarmer	Nature of Elusiveness	Why?
Scythians vs. Macedonians	Mobility	Horses
Parthians vs. Romans	Mobility	Horses
Seljuk Turks vs. Byzantines	Mobility	Less armor
Turks vs. Crusaders	Mobility	Less armor
Mongols vs. Eastern Europeans	Mobility	Horses, multiple mounts per man
Woodland Indians vs. U.S. Army	Mobility	Dispersed formations foraged off the land, carried fewer supplies
Napoleonic Corps vs. Austrians	Concealment	Nature of wooded terrain, indigenous support
Boers vs. British	Mobility and concealment	Horses, nature of terrain, indigenous support
U-boats vs. British Convoys	Concealment	Submersible U-boat, Schnorchel, countermeasures against ASDIC (burrowing in the mud in shallow water)
Somalis vs. U.S. Commandos	Concealment	The use of noncombatants, nature of urban terrain

known to cut the artery in their horses' neck and drink the blood while on the march. Napoleon's corps foraged off the land, taking what they needed from the countryside.[6] See Table 4.4.

---

[6]The last army to experience a revolutionary leap in logistics capability was *La Grande Armée* of Napoleon's day. Superior logistics was one of the secrets behind Napoleon's ability to outmaneuver his adversaries and rapidly concentrate his *corps d'armée* system. His ground forces enjoyed shorter logistical tails because they lived off the land during the march, rather than transporting all their supplies with them on the campaign. Because of the *levee en masse* and internalized discipline, French soldiers could be trusted to disperse and forage for supplies without taking the opportunity to desert their comrades. The greater dispersion and speed of Napoleon's corps allowed him to conduct war at the operational level.

Table 4.4

Logistics Problems and Solutions for Swarmers in the Past

Swarmer vs. Nonswarmer	Logistics Requirements	Logistics Solution
Scythians vs. Macedonians	Food and fodder, firewood	Operate in grazing terrain
Parthians vs. Romans	Food and fodder, firewood	Operate in grazing terrain
Seljuk Turks vs. Byzantines	Food and fodder, firewood	Operate in grazing terrain
Turks vs. Crusaders	Food and fodder, firewood	Operate in grazing terrain
Mongols vs. Eastern Europeans	Food and fodder, firewood	Operate in grazing terrain
Woodland Indians vs. U.S. Army	Food	Have indigenous support; keep campaign short
Napoleonic Corps vs. Austrians	Food and fodder, firewood, gunpowder, ammunition	Forage off the land in dispersed formations
Boers vs. British	Food and fodder, firewood, gunpowder, ammunition	Have indigenous support
U-boats vs. destroyers	Food, fuel, ammunition	Use surface resupply ships
Somalis vs. U.S. Commandos	Food, ammunition	Have indigenous support; keep campaign short

In pre-industrial cases, the logistics requirements were minuscule compared with the needs of a modern mechanized force. Before war became mechanized, food, fodder for the horses, and firewood for cooking were the critical supplies required for an army to march and fight. Mechanized armies with crew-served weapons demand ammunition and parts several orders of magnitude greater than those of the armies that were fielded before WWI.[7]

---

[7]The increasing mechanization of war since WWI has led to an exponential growth in the logistics demands of combat units. While crew-served weapons now expend vast quantities of ammunition, armored vehicles place greater and greater fuel demands on modern supply systems. Even during WWII, most successful armored break-

History can be used to identify the problems that swarmers encountered in the past. It can also highlight *by omission* new problems for present and future forces. For example, swarmers have never encountered minefields. If swarming is used in the future, will extensive mining prevent swarmers from dispersing and converging? Table 4.5 highlights these problems.

Table 4.5

Are Historical Constraints on Swarming Still a Problem?

Problem	Explanation	Present Solution Available?
No knockout blow	Lightly armed swarmer units had to wear down the nonswarmer army through the standoff fire of the bow, an attrition strategy without a knockout blow. Swarmer effectiveness is greatly enhanced when it is also capable of shock action. The Mongols had heavy cavalry to deliver *coup de grâce* after softening up the target.	Indirect fires may provide a knockout blow if the enemy is concentrated and located by sensors. Airpower (such as AC-130 gunships), MLRS ATACMS, field artillery, offshore Naval fire support, and space weapons can be used to provide indirect fire support.
Restrictive terrain	Noninfantry swarmers need maneuver space (roads, grazing lands, desert and open plains, the ocean, and the air). There are no cases of mounted swarmers operating in difficult terrain.	Light strike vehicles are capable of operating in most terrain. Dismounted swarmer teams are possible.
Logistics	Pre-mechanized age had relatively small logistics burden. Even so, either the swarmer had to be on the defense or a minor logistics RMA was needed.	A package of fixes: GPS delivery, air superiority, information-management improvements, etc.
Minefields	Nonexistent before WWI. Coastal sea mines used against U-boats.	Could be a problem. Support political ban?
Fratricide, disregard for high friendly casualties	Sometimes swarmers suffered very little (Boers), sometimes a lot (Somalis).	Improved situational awareness, precision engagement. Use direct-fire weapons that will not harm friendly forces on the opposite side of the target. Use nonlethal weapons.

throughs by either side on the Eastern Front were invariably brought to a halt by the lack of logistics support. Today, the 68-ton U.S. main battle tank requires refueling every 8 hours and can expend its limited basic load of ammunition in minutes of heavy contact.

A modern swarming doctrine demands superiority in many technological areas. Therefore, its ultimate feasibility will depend on the
benefits emerging from the revolution in communications and computers. Many benefits are already being realized. Several underway
governmental research and development efforts may be relevant to a
discussion of swarming, including the Army's Army XXI and AAN
work and the Marine Corps' Urban Warrior program.

The U.S. Army is already proceeding with its FORCE XXI modernization effort to eventually create "Army XXI." [8] Army XXI units will be
"digitized," communicating via a tactical internet and relying on a
network of communication systems to improve situational awareness. Concentration of fire will be more important than concentration of mass on the battlefield. Army XXI operations will be more
dispersed and nonlinear than in the past.

Planning also is under way for a modernized force called the Army
After Next (AAN), which will rely on weapons and technologies available after 2010. Preliminary studies by Training and Doctrine Command (TRADOC) on the likely characteristics of an AAN force have
tentatively concluded that greater dispersion of units, lighter-weight
vehicles, "air-mechanized" forces, and a more network-based organization are desirable. The current vision is of light battle forces that
can be air-mobiled across the theater using advanced airframes.
Units will maneuver in dispersed operations and rely in part on
"reachback" indirect weapons. [9] The light battle force essentially uses
a concept similar to swarming in that it waits for an enemy to
penetrate its defensive area, relies upon stealth to elude the enemy,
and unleashes an ambush of massive simultaneous fires from close
in and afar. [10]

---

[8] FORCE XXI is the Army's modernization process to field an Army for the twenty-first
century. The Army is upgrading many of its major weapon systems, but current modernization involves few new weapon systems. The central effort is the digitization of
the force—the application of information and communication technologies (drawn
for the most part from the commercial sector) to share real-time information with
every dismounted soldier, vehicle, aircraft, weapon, and sensor on the battlefield.

[9] *"Reachback" indirect weapons* are very long-range weapons deep in the rear.

[10] See unpublished RAND research by J. Matsumura, R. Steeb, T. Herbert, S.
Eisenhard, J. Gordon, M. Lees, and G. Halverson, "The Army After Next: Exploring New
Concepts and Technologies for the Light Battle Force."

The Marine Corps is exploring an "Urban Swarm" operational concept as part of its urban-warfare experimentation.[11] The Marines like to say that the urban swarm is similar to police tactics in emergency situations. Marine swarming calls for multiple squad-sized fire teams patrolling assigned areas, responding to crises, and calling for backup from other fire teams when necessary.[12] The Marine Corps After Next (MCAN) branch of the Marine Corps Warfighting Lab is also taking a look at a future force that will be dispersed, autonomous, adaptable, and small.

It is probable that technological advances associated with these efforts will alleviate some of the historical problems with swarming identified in this monograph. In the next chapter, this critical assumption—that future developments will make modern swarming feasible for U.S. light or medium forces—will be explored further with a discussion of swarm tactics, logistics, command, organization, and technology.

---

[11]The Marines conducted limited experiments on the East Coast as part of their Urban Warrior program, to learn and apply new tactics for military operations on urbanized terrain (MOUT). The entire effort culminated in a final Advanced Warfighting Experiment on the West Coast in spring 1999.

[12]George Seffers, "Marines Develop Concepts for Urban Battle Techniques," *Defense News*, January 12–18, 1998.

Chapter Five

TOWARD A SWARMING DOCTRINE?

Although the current FORCE XXI modernization effort is well under way and the Army hopes to complete the digitization of its first Army XXI Corps within a few years, the basic doctrine and division-based design of the ground force will not be changed radically. Any proposal (such as swarming) that calls for dismantling the Army division or the reinforced Marine battalion as the basic organizational building block will need to recognize that 60 years of doctrinal and institutional history present formidable obstacles. The service institutions themselves would have to change, as would joint doctrine.

For example, in *Breaking the Phalanx: A New Design for Landpower in the 21st Century* (Westport, CT: Praeger Publishers, 1997), Douglas MacGregor (of the Center for Strategic and International Studies in Washington, DC) proposes a new organization for the future army and argues that the division should be disestablished in favor of the brigade task force. MacGregor suggests that combined-arms "combat groups" will be effective, given the technology and missions of the future. His thought-provoking book provides detailed command, organization, and equipment tables that deserve serious consideration.[1] But even MacGregor did not stray too far in changes to equipment.[2]

---

[1] It is not clear that MacGregor's changes to doctrine and modernization are feasible in the short term (before Army After Next), nor is it clear that his combat groups would be effective across the threat environment, especially in peace operations. He probably did not include a large enough support structure.

[2] MacGregor lowered the number of tanks and infantry fighting vehicles required for his battle groups, but he did not replace them.

A doctrine based on swarming calls for more-radical changes in equipment and organization. Abrams tanks and Bradley fighting vehicles are not mobile enough for swarming operations, because they require major refueling and maintenance support. Tanks are designed for massing fires, not for dispersed operations with long-range fires.

Any doctrine that calls for such a drastic reduction in heavy vehicles will encounter a lot of resistance. A major shift in doctrine is risky. DoD is also currently spending billions every year to maintain a "warm" tracked-combat-vehicle industrial base.[3] The U.S. Army has bought about 8,000 Abrams tanks and 6,500 Bradleys since 1980. Plans are under way to field a future tank around the 2015–2020 timeframe.[4]

At the same time, the Army recognizes the need for a lighter force that can deploy rapidly and stand against heavy ground forces. That force does not exist. Because of their weight, U.S. heavy forces are not rapidly deployable.[5] DoD is already seeking a solution. Ongoing research initiatives such as the Rapid Force Projection Initiative (RFPI) Advanced Concept Technology Demonstration (ACTD) are investigating new technologies that enable light airborne forces to

---

[3]Recent government studies of the Tracked Combat Vehicle Industrial Base (Abrams and Bradley production) have estimated that, to keep the tank industrial base *warm*—capable of producing new tanks and expanding production within a short time (about 2 years)—the minimum necessary production activity would be to upgrade 120 Abrams tanks each year at the Lima, Ohio, tank facility (this is, in fact, what the Army currently does). In 1993, the procurement cost to maintain a warm production base was estimated at $650 billion. See Congressional Budget Office, "Alternatives for the US Tank Industrial Base," CBO Papers February 1993; Office of the Under Secretary of Defense, Industrial Capabilities and Assessments, *Industrial Assessment for Tracked Combat Vehicles*, October 1995.

[4]Both the Army and Congress are interested in preserving the current industrial base and maintaining an armor and ballistic structure production capability, but a debate exists over which tank-modernization path the Army should follow. The Army Science Board (ASB) argues for an "evolutionary" path to the next-generation tank: an Abrams block upgrade (M-1A4) between 2008 and 2020, which would ensure a warm tank industrial base. The Armor Center at Ft. Knox, Kentucky, favors a "leap ahead" approach to the Future Combat System (FCS), which precludes any M-1A2 production beyond 2003.

[5]Abrams tanks must be shipped to their destinations, a process that takes weeks. Once there, they usually need heavy-equipment transporters to move them far from the shoreline.

face advancing enemy armor.[6] In the fall of 1999, Army Chief of Staff General Eric Shinseki announced plans to redesign two combat brigades into a mobile force that can be rapidly deployed to any crisis spot in the world. Lighter alternatives to the Abrams tank and the Bradley IFV are currently being sought for this new "Strike Force."[7]

Because of these constraints, this monograph recommends that a doctrine based on swarming is more appropriate for future light or medium forces. History suggests that swarming works when an army possesses standoff capability, the ability to elude its opponents, and superior situational awareness. Ongoing technological development suggests that light Army or Marine units may enjoy these fundamental advantages in the future. Assuming that technological solutions are developed, a swarming doctrine may not only be feasible, it may be more appropriate than the currently organized division-based forces for certain light force missions. It is hoped that this survey has demonstrated that swarming has worked well in the past when certain advantages obtained; it is reasonable to assume that it can do so again if future forces enjoy those same advantages.

This discussion of historical swarming and the feasibility of using swarming in the future serves as another step in the process of proposal and debate about future U.S. military doctrine. The next step is to further flesh out the details of organizational design, the specifics of weapon platforms, and the feasibility of dispersed battlefield logistics without traditional lines of supply. Swarming scenarios have already been used in some high-level wargaming exercises.[8] Further

---

[6]RAND has a long history of exploration, analysis, and modeling of weapons and sensors for light forces, with both direct- and indirect-fire capabilities. One indirect-fire concept is the "hunter/standoff killer" concept, whereby assorted manned or unmanned hunters sense the enemy and communicate target coordinates back to either C2 nodes or indirect-fire assets (killers). See R. Steeb, J. Matsumura, T. G. Covington, T. J. Herbert, and S. Eisenhard, *Rapid Force Projection: Exploring New Technology Concepts for Light Airborne Forces*, Santa Monica, CA: RAND, DB-168-A/OSD, 1996.

[7]Ed Offley, "Fast Strike Force Being Developed at Fort Lewis," *Seattle Post-Intelligencer*, November 3, 1999.

[8]The Office of the Secretary of Defense (OSD) and U.S. Army Deputy Chief of Staff for Operations and Plans (DCSOPS) have conducted wargames that explore a swarming operational concept (at the Dominating Maneuver Game VI, 30 June–2 July 1997, U.S. Army War College, Carlisle Barracks, PA). Their view of swarming is that maneuver forces allow enemy forces to advance fairly unaware until they are attacked from all

study is needed to determine whether computer simulations, gaming, and testing might offer insights.[9] Eventually, an experimental force (EXFOR) could be created from an existing battalion, trained and equipped for swarming, and sent to a combat training center such as the Joint Readiness Training Center (JRTC) at Ft. Polk, Louisiana, for force-on-force training with an opposition force (OPFOR). Practical field experiments would help demonstrate whether swarming is feasible.

Since a future swarming doctrine is still very much a concept in progress, additional detail is offered here on the tactics, logistics, command and organization, and technology of a possible swarming doctrine. Limitations of and situations conducive to swarming are also described. This speculative discussion is based as much as possible on the historical conclusions from the ten cases.

TACTICS

Swarming can be conceptually broken into four stages: locate, converge, attack, and disperse. Swarming forces must be capable of sustainable pulsing: Swarm networks must be able to come together rapidly and stealthily on a target, then redisperse and be able to recombine for a new pulse.[10] It is important that swarm units con-

---

directions simultaneously. The swarm concept is built on the principles of complexity theory, and it assumes that blue units have to operate autonomously and adaptively according to the overall mission statement. The concept relies on a highly complex, artificial intelligence (AI)-assisted, theater-wide C4ISR architecture to coordinate fire support, information, and logistics. Swarm tactical maneuver units use precise, organic fire, information operations, and indirect strikes to cause enemy loss of cohesion and destruction. Swarming blue units operate among red units, striking exposed flanks and critical command and control (C2), combat support (CS), and combat service support (CSS) nodes in such a way that the enemy must constantly turn to multiple new threats emerging from constantly changing axes. Massing of fire occurs more often than massing of forces.

[9]The Center for Naval Analyses has already started computer simulation of swarming behavior by modeling combat as a complex, adaptive system with a set of simple, multi-agent "toy models" called ISAAC/EINSTein. These models assume that land combat is a complex adaptive system—essentially a nonlinear dynamic system composed of many interacting semiautonomous and hierarchically organized agents continuously adapting to a changing environment. Patterns of behavior may be observed from the decentralized and nonlinear local dynamics of the agent-based model.

[10]Arquilla and Ronfeldt, 1995, p. 465.

verge and attack simultaneously. Each individual swarm unit is vulnerable on its own, but if it is united in a concerted effort with other friendly units, overall lethality can be multiplied, because the phenomenon of the swarm effect is greater than the sum of its parts. Individual units or incompletely assembled groups are vulnerable to defeat in detail against the larger enemy force with its superior firepower and mass.[11]

Because of the increasing vulnerability of massed formations on the ground to airpower and WMD, the Dispersed Swarm maneuver is more appropriate for the future. More-dispersed operations are a natural response to the growing lethality of modern munitions.

Lightly armored and dispersed units must use elusiveness as a form of force protection. Lawrence of Arabia evoked the swarm philosophy of elusiveness when he compared the tactics of his Arab forces with those of his conventional Turkish opponents. Lawrence knew the Arabs needed a mobile force that would form "an influence, a thing invulnerable, intangible, without front or back, drifting about like a gas."[12]

In the past, some swarming units enjoyed what could be called "direct standoff fire," the capability to inflict damage on the enemy without receiving punishment in return, using weapons such as the composite bow or Mauser rifle. Today, swarm units can use indirect standoff weapons (both *organic*—carried on their persons—and *nonorganic*—a remote asset that has to be called to) such as missile-launched "brilliant" munitions or offshore naval platforms deep in

---

[11]For example, the Chechen tank killer teams that preyed upon lone Russian T-72 tanks on the streets of Grozny in 1995 accomplished their kills by arriving at (or attacking) the target at the same time. Piecemeal RPG attacks by teams arriving at different times might have been suppressed or defeated in turn.

[12]Lawrence of Arabia's guerrilla campaigning on the Arabian peninsula during WWI is similar to swarming. Lawrence's Arab forces did not swarm; they conducted hit-and-run attacks. Nevertheless, his tactics relied on superior mobility and intelligence just as swarm tactics do. Lawrence sought to avoid direct battle and exploit the immense open space of the desert to cut the Turkish lines of communication—a strategy he could afford to employ (there was no political pressure to defend cities) and one of the major differences between swarming and guerrilla warfare. The peculiar political conditions of Lawrence's Arabian campaign enabled him to employ a Fabian strategy whereby he could abandon major cities, whereas swarming does not. See Asprey, 1994, p. 184.

the rear. Because of this ability to apply force against a target with assets located far away, light units will potentially be much more lethal than their counterparts from the past. Swarm tactics aim to leverage this shift from direct-fire to indirect-fire weapons to some degree in order to improve the mobility of the individual unit on the ground and reduce its signature.

LOGISTICS

Before the proven WWII division structure will be dismantled, the following questions must be addressed: Which combat service support assets should be organic to swarm units and which assets should be prepositioned? To what extent is aerial resupply possible? How can information technology ease logistics demand?

Superior mobility will require substantial fuel resupply, barring some revolutionary development in chemical propulsion technology. Without the typical main support battalion that is normally present in a division area of operations, swarm units will require a new way to repair and maintain their vehicles.

Since swarming requires vehicles that are either mobile or stealthy, Abrams tanks and Bradley fighting vehicles are not ideal. New generations of light strike vehicles (LSVs)—which require far, far less logistics support, less fuel, and no heavy-equipment support vehicles[13]—may prove more practical.

A swarm unit could also use more indirect-fire assets rather than organic direct-fire weapons, limiting its ammunition load. It would also have a smaller tail of support personnel (compared with that of a division-based force), which would lower logistics demand.

Another way to possibly reduce the demand for supplies is to use *focused logistics*, the U.S. Army's operational concept to leverage

_____

[13]The Army's RFPI ACTD is looking at heavy HMMWVs to evaluate hunter/standoff killer operational techniques. The sensors of these light vehicles will include second-generation Forward-Looking Infrared Radar (FLIR) with embedded aided target recognition, acoustics, daylight TV, laser rangefinding, color digital maps, image compression/transmission, GPS, and secure communications. The goal is to transmit digital targeting reports plus imagery in near-real time to "killer" indirect-fire assets in the rear.

information technologies and thereby rapidly provide supplies such as food, fuel, equipment and ammunition.[14] The Marines call this *anticipatory logistics*, but it is basically the same thing. The notion here is that information-management systems allow constant visibility of all supplies to be maintained so that no unit need stockpile for emergencies. Smaller logistical tails result when ground combat units carry exactly those supplies they need, never more than necessary. Greater speed will help agile units maintain a faster tempo of operations. Considering both the frictions of war and the certainty of an adversary's trying to exploit dependence on such finely tuned logistics, how far the Army and Marines will be able to go is as yet unclear.

In short, the logistics problem of supplying widely dispersed units without traditional CSS battalions present is a difficult problem that will probably need to be addressed by a package of fixes. Additional fixes include the following:

- Spread the burden. Networked units can coordinate their supply needs, using situational awareness to transfer and share between units.

- Use common parts and systems. The organization of a swarming force will naturally be flat, with homogeneous unit types. Instead of ten battalions of Abrams and Bradleys, there might be 40 swarm units equipped with LSVs. With commonality of parts, there are fewer different types of systems to repair.

- Use precise aerial resupply when possible, including unmanned delivery systems such as GPS-guided parafoils.

---

[14]Focused logistics uses a Velocity Management approach to battlefield distribution, wherein the speed and control of logistics material is more important than the mass of stockpiles. By re-engineering logistics processes, Velocity Management can reduce the long material flows that help create massive stocks of supplies. Eliminating non–value-added activity (such as obtaining an extra signature from a middle manager) and maintaining in-transit visibility (or knowing where every logistics item is at all times) decreases the logistician's response time to warfighter demands. In the past, U.S. inventories have typically been large because warfighters hoarded supplies "just-in-case" the items they ordered either took too long to arrive or never showed up. Rather than "just-in-case," focused logistics seeks to respond to real-time battlefield demand and move in the direction of a "just-in-time" philosophy. Rapid response to the needs of dispersed maneuver units will provide logistics support in hours and days rather than weeks.

- Use prepositioned supply depots.

- Create combat service support units that operate with particular clusters of swarm units. These CSS units remain mobile but carry no combat capability and are devoted to support functions.

Alternative ways to treat and evacuate casualties, such as telemedicine, need to be perfected. If casualties cannot be air-evacuated, swarm units must either carry the wounded themselves or consolidate them at temporary field hospitals.

COMMAND AND ORGANIZATION

The organization of a standard armored division includes not just four mechanized and six armored battalions, but other division support units such as an engineer battalion, a signal battalion, a chemical company, a brigade of artillery, an air defense battalion, and the division support command.[15] All of these support units provide critical functions. Finding an alternative way to provide that support—if it is needed for a swarm unit—is one of the next steps of a serious consideration of swarming. However, it is beyond the scope of this monograph to detail what a table of organization and equipment for a swarm unit should look like.

Organization and command are directly related. Organization determines the number, position, and responsibilities of noncommissioned and commissioned officers. Whenever a force of any size is divided into many parts, the problem of coordination between units becomes more difficult. The complexity of the command problem grows with the number of units, the power and range of their weapons, the speed at which they move, and the space over which they operate.[16]

---

[15]Although the division structure has evolved somewhat since 1917, the major issue of Army force development has remained establishing the numbers and types of support units that should be in the echelons above division. As weapons have improved in lethality, the proportion of the army devoted to combat has decreased and the proportion devoted to support has increased.

[16] Van Creveld, 1985, p. 6.

Improved C4I technologies may provide part of the answer by increasing the supply of command, but a swarming doctrine will have to provide other ways to reduce the demand for command.[17] One way is to adopt a decentralized system of command in which orders flow from the bottom up rather than from the top down. The other way is to address training.

The extreme decentralization of a network organization with semi-autonomous units calls for the mission-order system of command (the German concept of *Auftragstaktik*). In the mission-order system, small-unit commanders are granted the freedom to deal with the local tactical situation on the spot while following the overall commander's *intent*.[18] Historically speaking, those armies that have allowed tactical commanders considerable latitude have been very successful. Roman centurions and military tribunes, Napoleon's marshals, and Mongol *toumen* commanders all demonstrated how the initiative of a subordinate leader initiative can minimize the complexity of hierarchical, top-down control.[19] Swarming would never work with a hierarchical command structure, because an extremely flat organization would place too much demand on the overall commander.

Achieving superior situational awareness may tempt higher-level commanders to exercise more control over tactical commanders on the scene—an urge that should be resisted. There is a natural tension between the decentralized system of *Auftragstaktik* and the very centralized command possibilities of the all-encompassing C4I system the U.S. Army is heading toward in its modernization effort. Even though high-level commanders may have unprecedented awareness of the battlefield, they should avoid micromanagement. Carl von Clausewitz's "friction" of war usually finds a way to ruin the best-laid

---

[17]For example, personnel could remain in swarm units for the duration of their enlistment in order to improve unit cohesion (and thereby reduce the demand on command).

[18]Joint doctrine already embraces the general philosophy of commander's intent and the mission-order. See Joint Pub 3-0, *Doctrine for Joint Operations*, 1 February 1995.

[19]Van Creveld, 1985, p. 270.

plan.[20] A decentralized command system would be more adaptable to friction caused by a loss of communications.

The character of swarming conflict—dispersed operations conducted by numerous small units in close coordination—will require that small-unit leaders assume high initiative and responsibility. For example, the squad is the basic swarm element in the Marine Corps "Urban Swarm." The Marine command and control concept states that the squad leader at the point of contact assumes the role of On-Scene Tactical Commander until relieved, operating within the intent of the overall commander but remaining in command even as higher-ranked officers from adjacent units arrive during the course of the battle. A higher-ranked commander can assume overall command only after becoming fully acquainted with the tactical situation. Implicit in this type of command and control arrangement is the requirement that junior leaders be capable of much higher levels of command and responsibility (squad leaders with the knowledge of a platoon leader of today).

Future enlisted personnel will most likely have to undergo more-extensive training than in the past. Recruits may need to be of higher quality to begin with (score in the upper half of the Armed Forces Qualification Test). All too often, the human-capital side of a new doctrine does not get the attention it deserves. Even today, recruiting and retaining high-quality personnel in the military are increasingly difficult.[21] The linkage between doctrine and personnel quality standards cannot be ignored.

---

[20]War is inherently a chaotic system where so many variables collide that a systematic breakdown of what actually occurs in any one battle is impossible. Clausewitz called this complexity and uncertainty the "friction" of war. *Friction* is used to represent all the unforeseen and uncontrollable factors of battle. In other words, friction corresponds to the factors that distinguish real war from war on paper. It includes the role of chance and how it slows down movement, sows confusion among various echelons of command, or makes something go wrong that has worked a hundred times before. See Carl von Clausewitz, *On War*, edited and translated by Michael Howard and Peter Paret, New York: Alfred A. Knopf, Inc., 1993.

[21]There are many reasons for this problem, including demographic trends, increased optempo and deployment overseas, a booming economy and a growing gap between civilian and military pay, and declining youth interest in enlistment while interest in attending college has grown (among high-quality youth).

TECHNOLOGY

The extent to which a swarming doctrine depends on superior technology is a key question. History demonstrates that technological advantage is *always* temporary. Technology alone will not suffice. Adversaries adapt to superior technology by either copying the technology or developing a countermeasure. In order for swarming to remain relevant, all the pieces of the RMA puzzle—doctrine, organization, and technology—must be fitted together properly.

That said, there are three critical functions that technology must enable for swarmer success: superior situational awareness, elusiveness (mobility and/or concealment), and direct and/or indirect standoff fire capability. The reader will no doubt note that these same capabilities, listed above, are already a major focus of the U.S. military's ongoing modernization effort.

Real-time situational awareness will require the integration of command and control systems, communication systems, and intelligence, surveillance, and reconnaissance systems.

The communication system for a dispersed tactical formation will have to be a mobile mesh communication network with high data throughput and survivability. Units must be capable of sharing information at all times, even in harsh electromagnetic environments.[22] Capture of nodes must not compromise system security. The Defense Advanced Research Projects Agency (DARPA) is already developing the kind of network communication systems essential for future swarmer units, especially dismounted swarmers.[23]

---

[22]There are many kinds of threats to the tactical internet. Radio-frequency bombs, conventional jamming, information-warfare attacks, even high-altitude electromagnetic pulse attacks are possible. See Sean Edwards, "The Threat of High Altitude Electromagnetic Pulse (HEMP) to FORCE XXI," *National Security Studies Quarterly*, Vol. III, Issue 4, Autumn 1997.

[23]One goal of DARPA's Small Unit Operations program is to develop a mobile wireless communication system for widely dispersed tactical units. This equipment will be capable of supporting a tactical internet based on dismounted-soldier and mounted-vehicle nodes without having to rely on a fixed ground infrastructure—essentially a "comm on the move" capability. The most promising type of system would be a mobile mesh network of communication nodes that are able to buffer, store and route packets of information. Such a system would be capable of non–line-of-sight transmission, a critical requirement for urban warfare. The military community must either develop such systems themselves or fund commercial enterprises, because the

Sophisticated ISR and target-acquisition capabilities will be essential to detecting and tracking enemy ground formations. Swarm units will need to rely on multiple layers of ground, airborne, and space-based sensors and a robust tactical internet. The key to effective fires on all battlefields will be accurate and dependable target location. Precision targeting systems must be able to locate and transmit target information quickly and accurately over reliable communications means in order to deliver indirect fire before it is too late.

Rapidly responsive indirect precision fires delivered by rockets, missiles, naval gunfire, or tactical air must be available. In most cases, a swarming operation will be a joint operation. Ideally, swarm units should possess both organic standoff precision munitions and a capability to call for indirect-fire assets.[24] Indirect munitions will need to be GPS-guided and capable of in-flight corrective maneuvering.[25]

---

marketplace is unlikely to produce a mobile wireless system with the necessary anti-jamming, security, and data-rate standards on it own. Commercial communications systems such as digital cellular systems are designed to achieve optimal spectral efficiency (bits per second per hertz), which is usually incompatible with good security characteristics such as low probability of detection. For an excellent study that addresses most of these issues, see Phillip M. Feldman, *Emerging Commercial Mobile Wireless Technology and Standards: Suitable for the Army?* Santa Monica, CA: RAND, MR-960-A, 1998.

[24]Clearly, a mobile light force would need a mix of direct and indirect fires. We should not expect small teams to get along without significant organic firepower, especially if the weather is cloudy or the terrain masks the movement of ground forces. For example, previous RAND work by Randy Steeb, John Matsumura, and colleagues concluded that current and near-future indirect-fire systems are not enough to protect a light airborne brigade against the assault of a division or more of enemy armor. See Steeb et al., 1996.

[25]Space restrictions do not allow a detailed discussion of the possible contributions of long-range indirect-fire assets. Clearly, one problem will be the short exposure time of targets moving between cover, traveling through urban areas, etc. Future systems may be able to detect targets at range, but the exposure time may be too short. For those standoff weapons that have 10, 20, or more minutes over target, the exposure time may be too short to engage the target. One way around this is to use loitering weapons or update-in-flight. For further information, see J. Matsumura, R. Steeb, T. J. Herbert, M. R. Lees, S. Eisenhard, and A. B. Stich, *Analytic Support to the Defense Science Board: Tactics and Technology for 21st Century Military Superiority*, Santa Monica, CA: RAND, DB-198-A, 1997.

Some kind of LSV will be needed to provide superior mobility for certain types of terrain.[26] This vehicle should probably also be light enough to be airlifted around the battlefield if necessary. Striking a balance between speed and survivability is the challenge. Vehicles light enough to be moved by helicopters, such as the 4-ton HMMWV, remain vulnerable because of their minimal armor; medium-weight vehicles, such as the LAV-25 (a 6 × 6 wheeled APC weighing about 14 tons), are top-heavy (although a fixed-wing aircraft, such as a C-130 or a C-17, could transport these to a theater).

It is the synergistic combination of these capabilities that matters. All parts of a "system of systems"[27] approach are mutually reinforcing and dependent. Standoff weapons need targeting data from ISR systems, ISR systems must be controlled with C2 systems, and communication systems provide the backbone for all other systems. No single technological capability will be a "silver bullet."

LIMITATIONS TO SWARMING

Battles are won by the careful meshing of one adversary's advantages with the other's weaknesses. Swarming is no exception. As with any tactic or strategy, swarming will not work against all types of opponents in all situations. It should be used in scenarios or missions for which it is most applicable. Even when swarm units have the advantages of superior situational awareness, elusiveness, and standoff fire, there are foreseeable missions and conditions for which swarming may not be ideal.

For example, the Massed Swarm maneuver used by the most-conventional armies in the past would present a problem today because the initial mass of troops that approached the battlefield (before they swarmed) would be vulnerable to modern munitions. The Dispersed Swarm maneuver is preferred, because it increases the survivability of future forces by allowing dispersed operations.

---

[26]In mountainous and other impassable terrain, dismounted units may be the only option.

[27]Admiral William A. Owens uses the term *system of systems* to represent a concept whereby weapons and systems from three technology areas—sensors, C4I, and precison guided munitions (PGMs)—will interact synergistically on future battlefields. See "The Emerging System of Systems," *Military Review*, May–June 1995, pp. 15–19.

Defensive swarming along a border or any area without maneuvering room could be a problem. If, prior to hostilities, swarm units have to defend border areas adjacent to the enemy, they are probably not well suited to providing a fixed, linear defense. The swarmer must either be allowed to preemptively swarm on the offense (see Figure 5.1) and cross into enemy territory first, or the attacker must be allowed to penetrate the swarmer's home territory in order to allow defensive swarmer attacks from all directions (see Figure 5.2).

Deliberate swarming attacks against fixed, defensive positions may not succeed when the defender has had time to fortify those positions and place extensive minefields.[28] A swarm attack that is channeled will fail. Heavily mined areas pose a problem for a swarming doctrine, which places so much emphasis on dispersion and maneuver.

If the enemy is an elusive guerrilla force in difficult terrain where vehicles cannot operate, only dismounted swarm units may be feasible. Dismounted swarm units will probably not have a direct standoff fire capability over their opponents (except perhaps at night).

SCENARIOS CONDUCIVE TO SWARMING

One small war rarely resembles another, so smaller wars tend to present more-unorthodox challenges to conventional powers. In general, the decentralized nature of swarm organizations offers added tactical flexibility, which may prove more advantageous in small-scale contingencies. Guerrilla campaigns present a special challenge for the conventional army; oftentimes, the dominant force must adjust its strategy and tactics to suit the nature of the enemy and the terrain.

---

[28]Extensive minefields might be a problem for swarm units with little logistics support, because they may not have the capability to clear and detect minefields quickly.

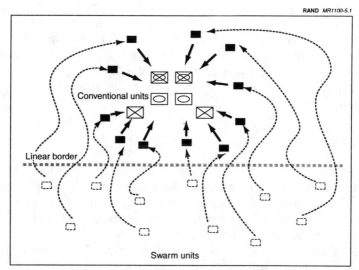

Figure 5.1—Offensive Swarming at a Border

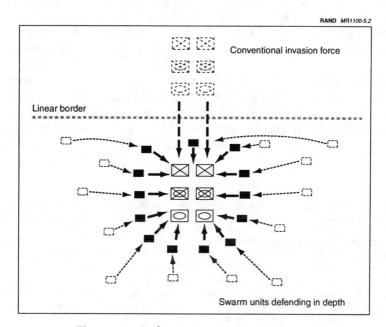

Figure 5.2—Defensive Swarming at a Border

In the changing post–Cold War environment, a network-based army that swarms may be better suited for certain missions. In fact, some of the very missions that U.S. military leaders foresee as being more likely in the future are conducive to swarming, including small-scale contingencies such as expeditionary (rapid force projection) operations, dispersed operations, counterinsurgencies, and peace operations.[29]

Rapid Force Projection

The problem of how to airlift heavy vehicles long distances will not be solved soon. As long as the United States must rely on relatively slow, oceangoing transport for its strategic lift, power-projection forces will not be able to rely on heavy weapons such as the 60-ton Abrams tanks at the beginning of a crisis. The most rapidly deployable units, such as the 75th Ranger regiment, are basically light infantry units without heavy, organic firepower.

Currently, the mission of rapid force projection is handled by light units such as the Marine Corps Expeditionary Units and Army airborne and air assault units, which lack tactical mobility and the capability to lay down heavy ordnance and so are vulnerable to enemy armor. Airborne units are capable of rapidly deploying (within a few days) and slowing down enemy armor, but not halting armor. What is needed is some sort of intermediate unit, in terms of mobility and mass, that can deploy between the arrival of the lightest forces and the arrival of heavy armored units some weeks later. The light nature of the hypothetical swarm unit seems ideally suited for this role.[30]

---

[29]AAN studies suggest that future land warfare will include contingencies that cover the spectrum of warfare, from operations other than war (which include peace operations and low-intensity operations) to high-intensity warfare, expeditionary operations, operations involving greater geographic scale, and operations with a greater probability of WMD use. See Lt. Gen. Edward G. Anderson III, "The New High Ground," *Armed Forces Journal International*, October 1997.

[30]The role in this case is rapid projection to defeat heavy ground forces. This monograph does not mean to imply that swarm units could supplant the United States' lightest forces for missions such as airfield seizures, raids, ambushes, and operations in the most difficult terrain such as mountains.

WMD Environments and Dispersed Operations

Given the increasing dominance of U.S. conventional power, future adversaries may conclude that the only effective way to fight the United States is to employ asymmetric strategies. One such asymmetric response is the use of chemical, biological, or even nuclear weapons against U.S. forces deployed in a regional crisis. Breaking a force into many smaller pieces and increasing the distance between maneuver units is one way to lower the vulnerability of U.S. forces.[31] Tactical dispersion provides force protection: The more dispersed a deployed ground force is, the less vulnerable it is to WMD and other destructive weapons.[32] Swarm units deployed in a network are ideally suited for this type of environment, assuming they would be trained to operate as an extremely dispersed force.

As Table 5.1 indicates, increasing battlefield dispersion is a natural historical trend, caused by the increasing lethality of weapons.

Future enemies may themselves also use dispersed operations to counter U.S. air superiority and artillery-delivered cluster munitions. As U.S. precision-fire assets continue to improve, adversaries may adopt dispersion as a tactical countermeasure for increasing their survivability.[33] The footprint of a typical U.S. deep-strike weapon such as the ATACMS Block IIA (a missile-delivered package of several brilliant anti-armor submunitions) is a fixed area; the fewer armored vehicles located in that footprint, the fewer kills are made. On the

---

[31]In the future, the emphasis will be even more concerned with dispersed operations as both sides improve their "see-it/kill-it" capabilities. In a "pop-up-warfare" environment, concealment is crucial to survivability. Dispersed forces can hide more easily.

[32]During the 1950s, the Army reorganized its divisions to fight on the atomic battlefield. It wanted to disperse American forces to minimize the damage from any single bomb. The "Pentomic" division consisted of units of five throughout its structure: 5 platoons per company, 5 battalions per brigade, etc. The Army viewed tactical nuclear weapons as a conventional warfighting option, and dispersion became the first imperative for its new Pentomic division. Tactics called for a rapid concentration, or massing, to defeat the enemy at the critical time and place, then a rapid and immediate dispersion. See A. J. Bacevich, *The Pentomic Era: The US Army Between Korea and Vietnam*, Washington, DC: National Defense University Press, 1986, p. 68.

[33]The recent lessons learned in Operation Allied Force in Kosovo in 1999 make it clear that concentrated masses of troops on the battlefield are highly vulnerable to weapons such as the satellite-guided Joint Direct Attack Munition.

future battlefield, concentrations of mass will be vulnerable to such deep-strike weapons. When the enemy does disperse into small groups of troops or vehicles to counter U.S. precision fires, a dispersed network of U.S. swarm units would be well positioned to swarm around clusters of enemy vehicles.

Table 5.1

Battlefield Dispersion from Antiquity to the Present

Area Occupied by Deployed Force, 100,000 Strong	Antiquity	Napoleonic Wars	U.S. Civil War	World War I	World War II	1973 Israeli- Arab War	Gulf War
Square kilometers	1	20.12	25.75	248	2,750	4,000	213,000
Front (km)	6.67	8.05	8.58	14	48	57	400
Depth (km)	0.15	2.50	3	17	57	70	533
Men per square km	100,000	4.79	3,883	404	36	25	2.34
Square meters per man	10	200	257.5	2,475	27,500	40,000	426,400

SOURCE: All figures except Gulf War column are from T. N. Dupuy, *The Evolution of Weapons and Warfare*, Indianapolis, IN: Bobbs-Merrill, 1980, p. 312. The area data for the Gulf War came from William G. Pagonis, *Moving Mountains: Lessons in Leadership and Logistics from the Gulf War*, Boston, MA: Harvard Business School Press, 1992; the rough number of 500,000 soldiers was used for the number deployed within this area.

Counterinsurgency Operations

We have already noted the similarities between guerrilla units and swarm units. The question is, Will swarm units be more effective than hierarchical division-based units in counterinsurgency (COIN) operations?[34] In the past, conventional armies have discarded their

[34]This does not imply that a network-based swarm force can replace Special Operations Force (SOF) personnel who are trained to raise guerrilla armies. In addition, Rangers, Civil Affairs, and PSYOPS units all provide unique functions that a regular Army swarm unit is unlikely to replicate.

heavy equipment and attempted to meet the enemy on equal terms, light infantry versus light infantry. But regular forces forfeit much of their technological superiority when they do so, sometimes absorbing higher casualties as a result. In contrast, swarm units deployed as a network will be more capable of finding and finishing elusive guerrillas. It takes elusiveness to counter elusiveness.

Intelligence-gathering is the heart of COIN operations.[35] The insurgent's knowledge of the local terrain and population is his greatest asset. He gains the support of the local population either by force or by popularity—can blend easily with the indigenous population, staying in safe houses and other hiding places.

For some time to come, U.S. space and air assets such as unmanned aerial vehicles, satellites, high-altitude aircraft and battle management aircraft like the Joint Surveillance and Target Attack Radar System (JSTARS) will be limited in their ability to detect small ground forces in difficult terrain such as mountains, heavy forests, cities, and jungles. Because of the limitations of surveillance technology and other uncontrollable factors such as inclement weather, the combination of light forces and rough terrain will probably remain impenetrable to our airborne sensors.

A swarm force can physically cover down over a geographic area,[36] and it is more likely to pick up battlefield intelligence. A network of swarm units dispersed over an area can perform such COIN missions as conducting frequent and random cordon search operations; establishing checkpoints that vary from location to location at random times; quickly reacting to suspected areas of insurgent activity when needed; and constantly gathering human intelligence.

Guerrilla armies usually operate as very small, light, and highly mobile units, are dispersed over a large area, and have no traditional flanks, rear, or line of communication. They must remain elusive to heavy conventional forces in order to survive. As Chairman Mao once said,

---

[35]O. Kent Strader, Captain, "Counterinsurgency in an Urban Environment," *Infantry,* January–February 1997.

[36]To *cover down* is to blanket or cover an area with numerous personnel. Units physically deploy in enough local areas that no area is left uninvestigated.

> When guerrillas engage a stronger enemy, they withdraw when he advances; harass him when he stops; strike him when he is weary; pursue him when he withdraws. In guerrilla strategy, the enemy's rear, flanks, and other vulnerable spots are his vital points and there he must be harassed, attacked, dispersed, exhausted and annihilated.[37]

Swarm units on the ground are natural sensors for detecting low-signature guerrilla units. Once a guerrilla unit is detected, all adjacent swarm units can seal off the area in which the guerrilla unit was last seen and converge toward that location. This is similar to what a traditional, hierarchical ground unit does today: It seals off the area and sweeps it in a linear, systematic fashion. But contracting a circle is much faster than sweeping a line across an area. Swarm units dispersed with their "ears to the ground" can ambush the ambushers.

Peacekeeping and Peacemaking Operations

Traditional combat divisions are not organized well for peace operations. Swarm units may be more effective at performing peace operations than hierarchical divisions because of their organization, ability to shape the environment, and decentralized command during operations in urban terrain.

Peace operations demand flexibility. Hypothetical swarm units make up a flatter, more flexible organization than the division-based army of today. Swarm units are more modular, and they can be reconfigured more easily than conventional units into task forces to support peace operations. The structure of the typical Army division-based task force today is not as well suited for peace operations, which emphasize policing, building, transporting, and facilitating rather than combat-arms functions. The current combat division does not contain all the unique personnel necessary to conduct these duties.[38]

---

[37]Asprey, 1994, p. 257.

[38]These missions demand a greater number of CS, CSS, and SOF forces. Because of this demand, task forces are sometimes drawn from the deployment of partial units and individual augmentees, rather than whole units. One might assume that because a hierarchical division-based force contains more CS and CSS personnel, it is more capable of forming peace operation task forces. However, most of the Military Occu-

A primary role of U.S. peacekeeping and peace enforcing units is to shape the environment with their actual physical presence. Whenever a peace mission calls for a patrolling presence of many small units dispersed over a large area, swarm units are ideally organized to adopt temporary duty of this sort. They can be trained to respond quickly to isolated incidents within their overall zone of control. Police forces around the world essentially use swarm tactics every day—for example, they swarm patrol cars to bank robberies in progress.[39]

Many peace operations are increasingly conducted in urban areas. The very nature of urban warfare requires decentralized control of assets, because communications capabilities are degraded, fields of fire and observation are limited, and mobility is reduced. The command system proposed in this monograph is designed specifically for decentralized command.

---

pational Specialty (MOS) augmentees in special demand for peace operations are in the Reserves. Activating these Reserve MOS augmentees requires a Presidential Selective Reserve Callup (PSRC), a mobilization step rarely taken for peace operations. In the absence of a PSRC, the demands of a major peace operation can strip critical MOS personnel such as military police from the rest of the active force that is not deployed.

[39]The bank robbery analogy is not perfect. Whereas police forces can afford to empty other areas of the city of police while responding to a crisis, military forces will not have that luxury because many "bank robberies" will be in progress.

Chapter Six

CONCLUSION

Swarming is not new. During the pre-gunpowder age, swarming armies enjoyed quite a bit of success on the Eurasian steppe and elsewhere; more recently, light infantry insurgents have fared well against conventional armies. The question is, Does a role exist for swarming today or in the future? History strongly suggests the answer is yes—if three capabilities can be achieved: superior situational awareness, standoff fire, and elusiveness. If emerging technology provides these capabilities, the United States could enter a watershed era of modern swarming that involves dispersed but integrated operations. Any doctrine of the future that relies on dispersed operations, such as the Army After Next or Urban Warrior, could benefit from a sustained research effort on swarming.

A radical departure from existing doctrine, a doctrine of swarming would require many issues to be worked out regarding tactics, logistics, command, and organization. Implementing such radical change, even on just a portion of U.S. ground forces, will require a careful yet bold plan that includes further research, gaming and simulation, and unit exercises before a prototypical swarming force is feasible. As well, the details of decentralized command and control will need to be worked out. A technological or doctrinal answer must be found for the logistics problems posed by a vehicle-based swarm force. But, because many of the likely conflict scenarios of the future—power-projection missions, counterinsurgencies, dispersed operations, and peace operations—appear to be conducive to a swarming doctrine, the investment will be worthwhile.

The limitations of swarming have already been noted. If dispersed swarmers are on the defense, invaders must be allowed to penetrate the swarmer's home territory before converging attacks can take place. Consequently, it is difficult for a dispersed swarmer army to defend a fixed line or border from penetration. Deliberate swarming attacks against fixed, defensive positions may not succeed when the defender has had time to fortify and channel swarm attacks with extensive minefields.

Swarming success in the past has also been highly dependent on terrain. Swarmers that were elusive because of their mobility relied on fairly unbroken terrain that could support large herds of horses. Swarmers that were elusive because of their ability to conceal themselves in dense forests or urban environments would never be able to operate in more open terrain. Yet swarmers have enjoyed marked success in the past, and they are likely to do so in the future if they are deployed with these limitations in mind.

The patterns and conclusions presented in this study are preliminary and are based on a carefully chosen yet limited sample. Indeed, further research on additional cases would help validate or complete the analysis begun here. Many other historical swarming examples remain, both from other battles between the belligerents examined in this study, and possible new cases such as the Battle of Britain in 1940, the defensive *Luftwaffe* tactics used over Germany late in World War II, the Chinese infantry tactics used in the Korean War of 1950–1952, the North American Indian Wars of the nineteenth century, and, more recently, the ongoing guerrilla war in southern Lebanon. Also, a closer look at battles between swarmers, such as Ayn Jalut and Homs, would explore how elusive forces fight equally elusive opponents. An analysis of all these additional cases would lead to stronger conclusions about what factors correlate with successful swarming.

REFERENCES

Alexander, Bevin, *The Future of Warfare*, New York: W. W. Norton & Company, 1995.

Anderson III, Lt. Gen. Edward G., "The New High Ground," *Armed Forces Journal International*, October 1997, pp. 66–70.

Arquilla, John, and David Ronfeldt, *In Athena's Camp: Preparing for Conflict in the Information Age*, RAND: Santa Monica, CA: MR-880-OSD/RC, 1997.

Arquilla, John, and David Ronfeldt, *The Advent of Netwar*, Santa Monica, CA: RAND, MR-789-OSD, 1996.

Arquilla, John, and David Ronfeldt, "Cyberwar Is Coming!" *Comparative Strategy*, Vol. 12, No. 2, Summer 1993, pp. 141–165.

Asprey, Robert B., *The War in the Shadows: The Guerrilla in History*, New York: William Morris and Company, 1994.

Atkinson, Rick, "Night of a Thousand Casualties; Battle Triggered U.S. Decision to Withdraw from Somalia," *Washington Post*, January 31, 1994a, p. A11.

Atkinson, Rick, "The Raid that Went Wrong; How an Elite U.S. Force Failed in Somalia," *Washington Post*, January 30, 1994b.

Bacevich, A. J., *The Pentomic Era: The US Army Between Korea and Vietnam*, Washington, DC: National Defense University Press, 1986.

Battin, Richard, "Early America's Bloodiest Battle," *The Early America Review*, Summer 1996.

Bowden, Mark, *Blackhawk Down: A Story of Modern War*, New York: Atlantic Monthly Press, 1999.

Bowden, Mark, *Philadelphia Inquirer* series on Somalia (http://www.philly.com/packages/somalia/graphics/2nov16 asp).

Chambers, James, *The Devil's Horsemen: The Mongol Invasion of Europe*, New York: Atheneum, 1979.

Chandler, David, *The Campaigns of Napoleon*, New York: The Macmillan Co., 1966.

Clausewitz, Carl von, *On War*, edited and translated by Michael Howard and Peter Paret, New York: Knopf, 1993.

Congressional Budget Office, "Alternatives for the US Tank Industrial Base," Washington, DC: CBO Papers, February 1993.

Contamine, Philippe, *War in the Middle Ages*, translated by Michael Jones, New York: Basil Blackwell, 1984.

Dougherty, Major Kevin, "Fixing the Enemy in Guerrilla Warfare," *Infantry*, March–June 1997, pp. 33–35.

Dupuy, R. E., and T. N. Dupuy, *The Encyclopedia of Military History from 3500 B.C. to the Present*, New York: Harper & Row, 1970.

Dupuy, T. N., *The Evolution of Weapons and Warfare*, Indianapolis, IN: Bobbs-Merrill, 1980.

Edwards, Sean, "The Threat of High Altitude Electromagnetic Pulse (HEMP) to FORCE XXI," *National Security Studies Quarterly*, Vol. III, Issue 4, Autumn 1997.

Eid, Leroy V., "American Indian Military Leadership: St. Clair's 1791 Defeat," *The Journal of Military History*, 1993, pp. 71–88.

Falls, Cyril, *The Art of War from the Age of Napoleon to the Present Day*, Oxford, England: Oxford University Press, 1961.

Feldman, Phillip M., *Emerging Commercial Mobile Wireless Technology and Standards: Suitable for the Army?* Santa Monica, CA: RAND, MR-960-A, 1998.

Fuller, John Frederick Charles, *The Generalship of Alexander the Great*, Piscataway, NJ: Rutgers University Press, 1960 (reprinted by DaCapo Press, Inc., New York, 1989).

Gore, Terry L., "The First Victory of the 1st Crusade: Dorylaeum, 1097 AD," *Military History*, Vol. 15, No. 2, June 1998.

Green, Peter, *Alexander of Macedon, 356–323 B.C.: A Historical Biography*, Berkeley: University of California Press, 1991.

Guerlac, Henry, and Marie Boas, "The Radar War Against the U-Boat," *Military Affairs*, Vol. 14, No. 2, Summer 1950.

Hezlet, Vice Admiral Sir Arthur, *The Submarine and Sea Power*, New York: Stein and Day, 1967.

Hildinger, Erik, "Mongol Invasion of Europe," *Military History*, June 1997.

Hildinger, Erik, *Warriors of the Steppe: A Military History of Central Asia, 500 B.C. to 1700 A. D.*, New York: Sarpedon, 1997.

Kaegi, Walter Emil, Jr., "The Contribution of Archery to the Turkish Conquest of Anatolia," *Speculum*, Vol. 39, No. 1, 1964.

Lehmann, Joseph H., *The First Boer War*, London: Jonathan Cape, 1972.

Leo VI, Emperor, *Tactica*, written around A.D. 900 (see Oman, 1998a, pp. 187–217).

Matsumura, J., R. Steeb, T. J. Herbert, M. R. Lees, S. Eisenhard, and A. B. Stich, *Analytic Support to the Defense Science Board: Tactics and Technology for 21st Century Military Superiority*, Santa Monica, CA: RAND, DB-198-A, 1997.

Maurice's Strategikon: Handbook of Byzantine Military Strategy, translated by George T. Dennis, Philadelphia: University of Pennsylvania Press, 1984.

Office of the Under Secretary of Defense, Industrial Capabilities and Assessments, *Industrial Assessment for Tracked Combat Vehicles*, Washington, DC: Congressional Budget Office, October 1995.

Offley, Ed, "Fast Strike Force Being Developed at Fort Lewis," *Seattle Post-Intelligencer*, November 3, 1999.

Oman, C.W.C., *The Art of War in the Middle Ages: A.D. 378–1515*, revised and edited by John Beeler, Ithaca, NY: Cornell University Press, 1953 (first published 1885).

Oman, C.W.C., *A History of the Art of War in the Middle Ages, Volume One: 378–1278 AD*, London: Greenhill Books, 1998a (first published 1924).

Oman, C.W.C., *A History of the Art of War in the Middle Ages, Volume Two: 1278–1485 AD*, London: Greenhill Books, 1998b (first published 1924).

Padfield, Peter, *War Beneath the Sea: Submarine Conflict During World War II*, New York: John Wiley & Sons, Inc., 1995.

Pagonis, William G., *Moving Mountains: Lessons in Leadership and Logistics from the Gulf War*, Boston, MA: Harvard Business School Press, 1992.

Plutarch, *Selected Lives from the Lives of the Noble Grecians and Romans, Volume One*, Paul Turner, ed., Fontwell: Centaur Press Limited, 1963.

Ransford, Oliver, *The Battle of Majuba Hill*, London: John Murray, 1967.

Rothenberg, Gunter E., *The Art of Warfare in the Age of Napoleon*, London: Batsford, 1977.

Seffers, George, "Marines Develop Concepts for Urban Battle Techniques," *Defense News*, January 12–18, 1998.

Steeb, R., J. Matsumura, T. G. Covington, T. J. Herbert, and S. Eisenhard, *Rapid Force Projection: Exploring New Technology Concepts for Light Airborne Forces*, Santa Monica, CA: RAND, DB-168-A/OSD, 1996.

Strader, Captain O. Kent, "Counterinsurgency in an Urban Environment," *Infantry*, January–February 1997.

Taw, Jennifer Morrison, David Persselin, and Maren Leed, *Meeting Peace Operations' Requirements While Maintaining MTW Readiness*, Santa Monica, CA: RAND, MR-921-A, 1998.

U.S. Department of Defense, *Counterguerrilla Operations*, Washington, DC: Department of the Army, FM 90-8, August 1986.

U.S. Department of Defense, *The Infantry Battalion*, Washington, DC: Department of the Army, FM 7-20, April 1992.

U.S. Department of Defense, *The Infantry Brigade*, Washington, DC: Department of the Army, FM 7-30, October 1995.

U.S. Department of Defense, *The Infantry Rifle Company*, Washington, DC: Department of the Army, FM 7-10, December 1990.

U.S. Department of Defense, *Operations*, Washington, DC: Department of the Army, FM 100-5, 1993.

U.S. Department of Defense, *Staff Organization and Operations*, Washington, DC: Department of the Army, FM 101-5, May 1997.

Van Creveld, Martin, *Command in War*, Cambridge, MA: Harvard University Press, 1985.

Van Creveld, Martin, *Technology and War: From 2000 BC to the Present*, New York: The Free Press, 1989.